Multi-Criteria Decision Models in Software Reliability

This book provides insights into contemporary issues and challenges in multi-criteria decision models. It is a useful guide for identifying, understanding and categorising multi-criteria decision models and ultimately for implementing the analysis for effective decision-making.

The use of multi-criteria decision models in software reliability engineering is a relatively new field of study, and this book collects all the latest methodologies, tools and techniques in one single volume. It covers model selection, assessment, resource allocation, release management, upgrade planning, open-source systems, bug tracking system management and defect prediction.

Multi-Criteria Decision Models in Software Reliability: Methods and Applications will cater to researchers, academicians, postgraduate students, software developers, software reliability engineers and IT managers.

Information Technology, Management and Operations Research Practices

Series Editors:
Vijender Kumar Solanki, Sandhya Makkar, and Shivani Agarwal

This new book series will encompass theoretical and applied books and will be aimed at researchers, doctoral students and industry practitioners to help in solving real-world problems. The books will help in various paradigms of management and operations. The books will discuss the concepts and emerging trends on society and businesses. The focus is to collate the recent advances in the field and take the readers on a journey that begins with understanding the buzz words such as employee engagement, employer branding, mathematics, operations and technology and how they can be applied in various aspects. It walks readers through engaging with policy formulation, business management and sustainable development through technological advances. It will provide a comprehensive discussion on the challenges, limitations and solutions to everyday problems such as how to use operations, management and technology to understand the value-based education system, health and global warming and real-time business challenges. The book series will bring together some of the top experts in the field throughout the world who will contribute their knowledge regarding different formulations and models. The aim is to provide the concepts of related technologies and novel findings to an audience that incorporates specialists, researchers, graduate students, designers, experts and engineers who are occupied with research in technology-, operations- and management-related issues.

For more information about this series, please visit: https://www.routledge.com/Information-Technology-Management-and-Operations-Research-Practices/book-series/CRCITMORP

Multi-Criteria Decision Models in Software Reliability

Methods and Applications

Edited by
Ashish Mishra, Nguyen Thi Dieu Linh,
Manish Bhardwaj, and Carla M. A. Pinto

CRC Press
Taylor & Francis Group
Boca Raton London New York

CRC Press is an imprint of the
Taylor & Francis Group, an **informa** business

First edition published 2023
by CRC Press
6000 Broken Sound Parkway NW, Suite 300, Boca Raton, FL 33487-2742

and by CRC Press
4 Park Square, Milton Park, Abingdon, Oxon, OX14 4RN

CRC Press is an imprint of Taylor & Francis Group, LLC

Library of Congress Cataloging-in-Publication Data
Names: Mishra, Ashish (Ashish Kumar), editor. | Dieu Linh, Nguyen Thi,
editor. | Bhardwaj, Manish (Professor of computer science and
engineering), editor. | Pinto, Carla M. A. (Computer scientist), editor.
Title: Multi-criteria decision models in software reliability : methods and
applications / edited by Ashish Mishra, Nguyen Thi Dieu Linh, Manish
Bhardwaj, and Carla M.A. Pinto.
Description: First edition. | Boca Raton : CRC Press, 2023. |
Includes bibliographical references.
Identifiers: LCCN 2022022574 (print) | LCCN 2022022575 (ebook) |
ISBN 9780367408824 (hardback) | ISBN 9781032342542 (paperback) |
ISBN 9780367816414 (ebook)
Subjects: LCSH: Computer software—Reliability—Mathematical models. |
Computer software—Development—Decision making. | Multiple criteria
decision making.
Classification: LCC QA76.76.R44 M85 2023 (print) | LCC QA76.76.R44
(ebook) | DDC 005—dc23/eng/20220915
LC record available at https://lccn.loc.gov/2022022574
LC ebook record available at https://lccn.loc.gov/2022022575

ISBN: 978-0-367-40882-4 (hbk)
ISBN: 978-1-032-34254-2 (pbk)
ISBN: 978-0-367-81641-4 (ebk)

DOI: 10.1201/9780367816414

Typeset in Times
by codeMantra

Contents

Preface

It is a matter of pleasure for us to put forth the book titled, *Multi-Criteria Decision Models in Software Reliability: Methods and Applications*. In the present era, software reliability plays a vital role in solving different kinds of problems and providing promising solutions in digital world. Because of the increase in digitalisation in today's lifestyle and each and every service to make the life easier, good software interfaces are required. Due to the increase in the usability and dependency on software, one important feature matters a lot, that is software reliability. The success of incorporation of the heavy software in the system works only with reliability feature. Such reliability depends upon different criteria and the deployed environment. It does not always relate to one or two factors, but it depends upon various factors such as physical or virtual.

This book explores various factors and criteria within different chapters related to reliability and decision-making steps. These aspects make decision-making approaches more powerful, reliable and efficient. The above-mentioned characteristics make the software reliability approaches more suitable and competent for decision-making systems. Nowadays, machine learning is incorporated in each and every field of engineering to make the automated system for better decision-making solutions. This kind of system provides the efficient decision in less time. Medical science and engineering have been using various medical systems such as medical imaging devices, medical testing devices and medical information systems. In order to analyse such big data efficiency, image processing, signal processing and data mining play important roles for computer-aided diagnosis and monitoring.

Decision-making in the medical field is a very important part because it is directly related to human life, so monitoring and diagnosis software should be reliable enough to provide the correct reports. This book will enable the reader to appreciate the applications of multi-criteria decision models in software reliability and their different methods used in various fields according the field criteria.

CHAPTER 1

This chapter focuses on building an item-item recommender system using collaborative filtering. The proposed model uses the well-known MovieLens dataset and also uses the concept of Bayesian average for evaluating movie popularity. In order to deal with the problem of sparsity, our proposed model builds compressed sparse row (CSR) matrix. This chapter uses machine learning approach using K-nearest neighbours for recommending movies based on similarity.

CHAPTER 2

This chapter focuses on the examination of relevant literature and provides a conceptual framework that explains the role of machine learning and profound learning in the development of intelligent (artificial) beings.

CHAPTER 3

This chapter reviews the various classifications used to predict software defects using software measurements in the literature. In this chapter, a detailed analysis of application of data mining and machine learning approaches used for software quality, defect and quality analysis is presented.

CHAPTER 4

This chapter analyses the types of ambiguities that arise due to poor management of requirement engineering and how it affects software quality and customer satisfaction. Moreover, it discusses the challenges an enterprise faces when, in prototype model, new feature are added continuously based on business requirements.

CHAPTER 5

This chapter describes the integration of multi-criteria decision making (MCDM)-based fuzzy analytic hierarchy process (FAHP) and fuzzy Technique for Order Preference by Similarity to Ideal Solution (FTOPSIS) methods that are applied for the formation or selection of best group of programmers.

CHAPTER 6

This chapter intends to use one of the unknown yet powerful machine learning algorithms, MCDM, to foresee the presence of heart disease in a person more accurately in order to save more lives by detecting and treating the patient before any major issue.

CHAPTER 7

In this chapter, the classification of software reliability models (SRMs) is studied on the basis of effective and efficient quality of SR models and obtains software faults with categorisation of vast variety of available software.

CHAPTER 8

This chapter provides a detailed study of different types of reliability models, which are responsible for the software reliability measurements. As every model has different criteria, so no single model is perfect. It also provides information about software quality improvement.

CHAPTER 9

This chapter shows the comparison of different techniques to resolve vulnerabilities using different multi-criteria decision analysis (MCDA) methods. The MCDM saves and sorts the list of criteria affecting the environments.

CHAPTER 10

This chapter describes and gives possible approaches for the safety assessment of AI systems. The AI system to integrate safety level needs and used for probabilistic failure behaviour for the dangerous part of the random budget for failure relevant in AI system.

CHAPTER 11

In this chapter, a step-by-step model for the FDP and FCP is proposed based on the ANN. The test initiative is taken into account as it has a strong impact on the error detection and correction process.

CHAPTER 12

In this chapter, various MCDM methodologies are studied with different perfor-mance parameters along with the new methodology FMCDM and its applications. The new methodology is compared with the traditional methodologies.

CHAPTER 13

In this chapter, to extend the capabilities of large-scale application and fix any faults detected during operation, software systems with optimisation help in selecting new techniques constantly for improving the next release sequence of plan, which is a huge challenge for firms developing or managing such vast and sophisticated systems.

CHAPTER 14

In this chapter, modelling data are evaluated with a deep neural network algorithm that is created expressly to predict the amount of faults, and the fault-free software system is finalised.

CHAPTER 15

This chapter reviews the recent technologies and uses deep learning mechanisms to detect vulnerabilities. It shows how they apply state-to-state neural techniques that are helpful for capturing probable vulnerable codes and patterns. It also provides complete reviews of the visions, concepts and ideas of the game modifiers for their field of interest.

We sincerely thank Ms. Erin Harris, Senior Editorial Assistant, CRC Press/Taylor & Francis Group, for giving us an opportunity to convene this book in her esteemed publishing house and for their kind cooperation in completion of this book, and Dr. Vijender Kr. Solanki, Sandhya Makkar and Shivani Agarwal, Series Editors in IT, Management and Operation Research. We thank our esteemed authors for having shown confidence in this book and considering it as a platform to showcase and share their original research work. We would also wish to thank the authors whose papers were not published in this book, probably because of minor shortcomings.

Editors

Dr. Ashish Mishra is currently working as a Professor in the Department of Computer Science and Engineering, Gyan Ganga Institute of Technology and Sciences, Jabalpur [M.P], India.

He is a qualified individual with around 19 years of expertise in teaching and R&D with specialisation in Computer Science Engineering. He completed B.E., M.Tech. and MBA. He received his Ph.D. degree from AISECT University, Bhopal, India. He has been a part of various seminars, webinars, paper presentations, research paper reviews and conferences as co-convener, Member of Organizing Committee, Member of Advisory Committee and Member of Technical Committee, and he has contributed to organising INSPIRE Science Internship Camp. He is a Senior Member of IEEE, Life Member of CSI and Secretary CSI Jabalpur Chapter. He has published many research papers in reputed journals and conferences. He also has papers in Springer and IEEE conferences. He is also a reviewer and Session Chair, Keynote Speaker of IEEE, Springer international conferences, CSNT-2015, CICN-2016, CICN2017, INDIACom-2019, ICICC-CONF 2019, ICICC-CONF 2020 and ICICC-CONF 2021. His research interests include IoT, data mining, cloud computing, image processing and knowledge-based systems. He published 30 patents in Intellectual Property India. He has published 8 books in the area of data mining, image processing and artificial intelligence.

Dr. Nguyen Thi Dieu Linh is currently working as a Dy. Head of Science and Technology Department, Hanoi University of Industry, Vietnam (HaUI). She received her Ph.D. in Information and Communication Engineering from Harbin Institute of Technology, Harbin, China. She has more than 19 years of academic experience in electronics, IoT, telecommunication, big data and artificial intelligence. She has published more than 30 research articles in national and international journals, books and conference proceedings. She is a reviewer for Information Technology Journal, Mobile Networks and Applications Journal and some international conferences. Now, she is an editor for some books such as *Artificial Intelligence Trends for Data Analytics Using Machine Learning and Deep Learning Approaches*; *Distributed Artificial Intelligence: A Modern Approach* published by Taylor & Francis Group, LLC; and *Data Science and Medical Informatics in Healthcare Technologies* published by Springer. Otherwise, she is an editor of International Journal of Hyperconnectivity and the Internet of Things (IJHIoT) IGI-Global, the USA.

Dr. Manish Bhardwaj is currently working as a Research Assistant Professor in the Department of Computer Science and Engineering, KIET Group of Institutions, Muradnagar, Ghaziabad, India. He is a qualified individual with around 11 years of expertise in teaching and R&D with specialisation in Computer Science Engineering. He received his Ph.D. degree from Dr. Abdul Kalam Technical University (AKTU), Lucknow, India. He completed M.Tech. (Computer Science & Engineering) from SRM University, Chennai (Gold Medalist, received award from former central health minister Mr. Gulam Nabi Azad). He is contributing to the scientific community by his enormous academics and research works in the areas of computer science, simulations, mobile ad hoc network protocols and wireless sensor networks. He has published nearly 60 Research Papers in various international journals/conferences. He has also taken part in nearly 150 international conferences and journals as General Chair, International Scientific Committee Members/Reviewer (SCOPUS index journals and conferences) and Editorial Board Member/Reviewer in reputed journals such as IEEE and Springer. He has contributed 1 book as an Editor and 8 book chapters in various renowned publications such as CRC Press and IGI Global. He has nearly 14 patents (ten national + four international).

Dr. Carla M.A. Pinto is a Coordinating Professor in the School of Engineering at Polytechnic of Porto, Portugal. Her main research topic is epidemiology, in particular Mathematical Epidemiology. She is interested in mathematical challenges and their role in providing advice on public health policies. Mrs. Pinto is trained in Nonlinear Dynamics, Bifurcation Theory. Previous research included the analysis of Central Pattern Generators for Animal and Robot Locomotion, coupled cell networks, and neuron-like equations (Hodgkin-Huxley equations, Fitz-Hugh Nagumo, and Morris-Lecar). She is an Associate Editor of international journals with a high impact factor. She is the Guest Editor of several books. She has published more than 100 articles. Her h-index is 20 and she has over 1700 citations.

Contributors

Aakriti
Bharati Vidyapeeth's College of
Engineering
New Delhi, India

D. Akila
Department of Computer Applications
Saveetha College of Liberal Arts and
Sciences
SIMATS deemed to be University
Chennai, India

Isha Bansal
Bharati Vidyapeeth's College of
Engineering
New Delhi, India

L. Bhagyalakshmi
Rajalakshmi Engineering College
Chennai, India

S. K. Bharadwaj
Madhav Institute of Technology &
Science
Gwalior, India

Manish Bhardwaj
KIET Group of Institutions, Delhi-NCR
Ghaziabad, India

Korhan Cengiz
University of Fujairah
Fujairah, UAE

Deviprtiya
Vels Institute of Science, Technology
and Advanced Studies
Chennai, India

V.R. Elangovan
Agurchand Manmull Jain College
Chennai, India

Kartik Gupta
Bharati Vidyapeeth's College of
Engineering
New Delhi, India

Sardar M. N. Islam
ISILC
Victoria University
Melbourne, Australia

Rachna Jain
Bhagwan Parshuram Institute of
Technology
Delhi, India

S. Jeyalaksshmi
Vels Institute of Science, Technology
and Advanced Studies
Chennai, India

Padmaja Joshi
CDAC
Mumbai, India

Rachana Kamble
Technocrats Institute of Technology
Bhopal, India

Aarti M. Karande
Sardar Patel Institute of Technology
Mumbai, India

Devansh Kashyap
Kalinga Institute of Industrial
Technology
Bhubaneswar, India

Shreyansh Keshri
Kalinga School of Management
Bhubaneswar, Odisha, India

Hassan Raza Mahmood
FAST NUCES Chiniot-Faisalabad
    Campus
Chiniot-Faisalabad, Pakistan

Ashish Mishra
Gyan Ganga Institute of Technology
    and Sciences
Jabalpur, India

D. K. Mishra
Madhav Institute of Technology &
    Science
Gwalior, India

Jyoti Mishra
Gyan Ganga Institute of Technology
    and Sciences
Jabalpur, India

Nishchol Mishra
RGPV
Bhopal, India

Neelu Nihalani
RGPV
Bhopal, India

Samad Noeiaghdam
Irkutsk National Research Technical
    University
Irkutsk, Russia
and
South Ural State University
Chelyabinsk, Russia

D. Padmapriya
Vels Institute of Science, Technology
    and Advanced Studies
Chennai, India

Vishal Paranjape
RGPV
Bhopal, India

Piramu Prithika
Vels Institute of Science, Technology
    and Advanced Studies
Chennai, India

Prashant Richhariya
Technocrats Group of Institutions
Bhopal, India

Harish K. Shakya
Amity University
Gwalior, India

Aditi Sharma
Parul University
Vadodara, Gujarat, India

Saurabh Sharma
Amity University
Gwalior, India

Vineet Sharma
KIET Group of Institutions, Delhi-NCR
Ghaziabad, India

Himanshu Shekhar
Hindustan Institute of Technology and
    Science
Chennai, India

Vikas Shinde
Madhav Institute of Technology &
    Science
Gwalior, India

Rajeev Shrivastava
Princeton Institute of Engineering &
    Technology for Women
Hyderabad, India

Ragini Shukla
Dr. C. V. Raman University
Chhattisgarh, India

Shweta Singh
KIET Group of Institutions, Delhi-NCR
Ghaziabad, India

Shubham Singh
Galgotias University
Greater Noida, India

Anurag Sinha
Department of Computer Science
IGNOU
New Delhi, India

Anita Soni
IES University
Bhopal, India

Sanjay Kumar Suman
St. Martin's Engineering College
Hyderabad, India

Kshitij Tandon
Jaypee University of Engineering and
   Technology
Guna, India

Narina Thakur
Bhagwan Parshuram Institute of
   Technology
Delhi, India

1 Enhancing Software Reliability by Evaluating Prediction Accuracy of CBF Algorithm Using Machine Learning

*Vishal Paranjape, Neelu Nihalani
and Nishchol Mishra*
RGPV

CONTENTS

DOI: 10.1201/9780367816414-1

1.1 INTRODUCTION

A vital factor affecting system reliability is software reliability. Alternatively, it is described as the likelihood of software being successfully executed for a particular instant of time. Several techniques were proposed for determining the software's reliability. A particular task is fulfilled by a software system in a particular environment for predefined number of input cases is termed as software reliability. A very important connection to software reliability is software quality, comprising functionality, usability, performance, etc. Software quality hinders the growth of software reliability. It is difficult to reach certain level of reliability with any system with a complexity. The machine learning approach guarantees to predict accurate solution to a given problem and therefore is a promising approach for ensuring software reliability. Today, machine learning approaches are used in a number of applications; one of the most used approaches is recommender systems where a user is being recommended items on the basis of his/her purchasing history of buying habits. A number of applications such as e-commerce, movies recommendation and social networking such as Facebook make use of recommender systems.

The entire chapter is divided into the following sections: Section 1.2 deals with the background details. Section 1.3 presents the ML techniques and methodology used for reliability assessment in our proposed work. The experimental set-up is discussed in Section 1.4. Results are represented in Section 1.5. Section 1.6 concludes the chapter.

1.2 BACKGROUND DETAILS & RELATED WORK

1.2.1 SOFTWARE RELIABILITY

An important feature for enhancing software quality is ensuring software reliability dealing with the bugs present in the system [1]. Fault in code is the major reason for failure in the system. Analytical models are used to measure the reliability of software termed as software reliability growth models (SRGMs) [2,3].

1.2.2 CRITERION TO MEASURE PERFORMANCE OF SGRM

Past research presented several techniques to acquire software reliability, but to access it and estimate mean time to failure (MTTF), we use a mathematical model called SRGM. There are two categories of SRGMs on the basis of nature of process:

1. Times between failures models
2. Fault count models.

Some well-known SRGMs are Goel-Okumoto, Musa-Okumoto, Jelinski-Moranda, etc. For deciding reliability level and to stop testing, we use these models [4].

For evaluating the performance of various models, we use several criteria such as root-mean-square error (RMSE), mean absolute error (MAE), average error (AE), and normalised root-mean-square error (NRMSE). Our proposed model uses only RMSE and MAE approach for evaluating the performance. The mathematical equations for the above-mentioned techniques are given below.

$$RMSE = \sqrt{\frac{\sum_{i=1}^{N}(x_i - \hat{x}_i)^2}{N}} \quad (1.1)$$

where
  i = Variable
  N = Number of non-missing data points
  x_i = Actual rating
  $\hat{x}_i$ = Predicted rating.

$$MAE = \frac{\sum_{i=1}^{N}|(p_i(f) - a_i(f))|}{N} \quad (1.2)$$

$$NRMSE = \frac{\sqrt{\sum_{i=1}^{k}(p_i(f) - a_i(f))^2}}{\sum_{i=1}^{k}p_i(f)^2} \quad (1.3)$$

where
  k = Number of failures
  $a_i(f)$ = Number of actual failures
  $p_i(f)$ = Number of predicted failures.

1.3 MACHINE LEARNING: A BRIEF OVERVIEW

A technique that is capable of learning from training data and predicting results is called machine learning. Broadly, we classify machine learning into four categories, which are discussed in the next section. Further, subcategorisation of the different types of ML is depicted in Figure 1.1 below. Under uncertainty, this technique plays a vital role in prediction and decision-making. On the basis of type of data and questionnaire being asked, different taxonomies of ML are available, which classifies machine learning. The classification of ML is given in Figure 1.1.

1.3.1 SUPERVISED LEARNING

In this method, we use labelled data with the help of which we train our model. In other words, we can say the learning that takes place in the presence of a supervisor is called supervised learning. The major part of this type of learning includes mapping function, which maps I/P variable (X) with the O/P variable (Y).

$$Y = f(X)$$

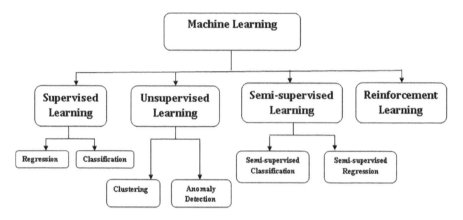

FIGURE 1.1 Categories of machine learning.

Supervision for model training is the main step involved in supervised learning. It can be simulated with the fact that proper learning takes place in the presence of a teacher or mentor in school. Two problems come in this category: **classification** and **regression**.

1. **Classification Models**: The problems in which output variables can be classified as "Yes" or "No", or "Pass" or "Fail" are categorised as classification models. In order to predict data category, we use these models. These can be binary classification or multiclass classification models. Some well-known examples for classification models that are deployed are spam filtering in emails, churn prediction, etc.
2. **Regression Models**: Whenever the output is predicted based on the previous data, we use the concept of regression models, for example house rent prediction. Linear, polynomial, ridge and logistic regression are some of the more familiar regression algorithms.

Regression problems are all about predicting $f\%$ for a quantitative response, such as blood pressure and temperature. For prediction, many ML algorithms are available, ranging from simple linear regression (LR) [5] and polynomial response surface (PRS) [6] to more complex support vector regression (SVR) [7], decision tree regression (DTR) [8], and random forest regression (RFR) [9]. By accurately quantifying uncertainty in regression problems, we use some machine learning (ML) models [10,11]. DNNs are more reliable than conventional ML equivalents and are effective in controlling the overfitting issue [12] (Figure 1.2).

1.3.2 UNSUPERVISED LEARNING

The learning that takes place in the absence of a supervisor is called unsupervised learning; in this type of learning, we do not have labelled data. This technique does not provide any training data. A large volume of data is fed to the machine for

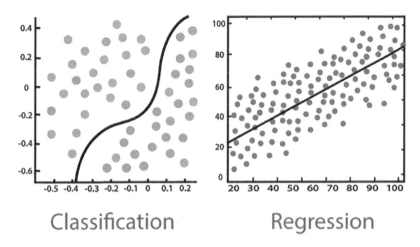

FIGURE 1.2 Classification and regression model.

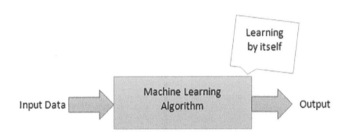

FIGURE 1.3 Unsupervised learning.

developing model and patterns, and on the basis of this learning, the model is fed with the testing data so as to provide efficient predictions. In unsupervised learning, there are no defined outcomes; moreover, it determines whatever different or interesting patterns exist in a given data set. Recommender system is basically based on the concept of unsupervised learning where we use several algorithms such as k-means clustering and k-nearest neighbours (Figure 1.3).

1.3.2.1 Categorisation of Unsupervised Machine Learning

1. Of all the learning methods, clustering is an important unsupervised learning method. Organising unlabelled data into similar groups is the main task of clustering technique. Therefore, collection of similar data items is called clustering. Grouping of similar data points into cluster and finding similar data points is the main goal of clustering.
2. The technique of identification of rare items or events differing from majority of data is called anomaly detection. Since anomalies or outliers are suspicious, generally we look for them. Bank fraud and medical error detection generally uses anomaly detection techniques.

1.3.3 SEMI-SUPERVISED LEARNING

A technique comprising of mix up of labelled data and unlabelled data during the phase of training is called semi-supervised learning. In this technique, first, the model is trained with the training data and then it is fed with the testing data to get the predictions.

To produce improvement and accuracy in learning, we use unlabelled data. A skilled human agent is required for acquiring labelled data for a learning problem or a physical experiment. It is relatively inexpensive to acquire unlabelled data.

A text document classifier is an example of this type of learning. It is so because it is not time efficient to have a person read the entire document. So, with the help of labelled text it becomes easy to classify labelled text with unlabelled (Figure 1.4).

1.3.4 REINFORCEMENT LEARNING

An interactive environment using hit and trial is learning which comes under the category of reinforcement learning (RL) and is an ML technique. Mapping between input and output is provided by both supervised and reinforcement learning where we give feedback to the agent. These feedbacks are of two types: Whenever there is a

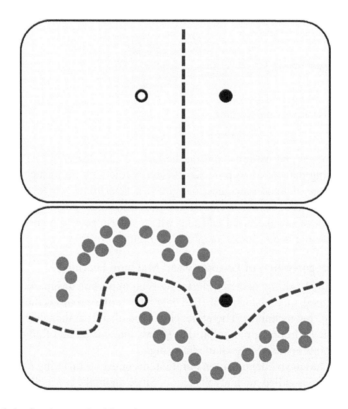

FIGURE 1.4 Semi-supervised learning.

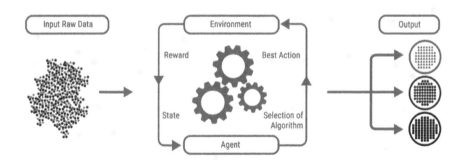

FIGURE 1.5 Reinforcement learning.

positive reward, then that type of performance is repeated, while if there is negative impact of a work, then it is avoided (Figure 1.5).

1.3.4.1 Algorithms Used in Machine Learning

Some commonly used machine learning algorithms are discussed below:

1. Linear Regression

   This technique estimates the exact values, for example total sales prediction and cost of houses, on the basis of continuous variables. The best line is fitted to depict the relationship between two variables. The line is also called regression line shown by the linear equation

$$Z = m * X + c$$

   where Z is dependent on the values of X and c, and m is the slope.

   For example, if we give an assignment to a student studying in fifth class to separate people according to their weight, then he on the basis of his skills will arrange people and separate them on the basis of their height and weight to classify them just by visualisation. This is a real-life application seen for linear regression. Figure 1.6 given below depicts a simple linear regression.

2. Logistic Regression

   As many a time we get confused by the name regression, whereas in real, it is a classification algorithm. Discrete values comprising values such as 0/1, yes/no and true/false are estimated by logistic regression. The probability of occurrence of event is predicted by fitting data. As this method is basically based on probability, its value generally lies between 0 and 1 (Figure 1.7).

3. Decision Tree

   A well-known algorithm used for classification problems is decision tree. Here, the entire population is split into two or more homogenous sets. In the diagram depicted below, we can see how a decision tree works. For

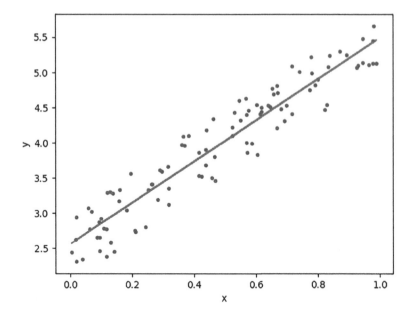

FIGURE 1.6   Linear regression.

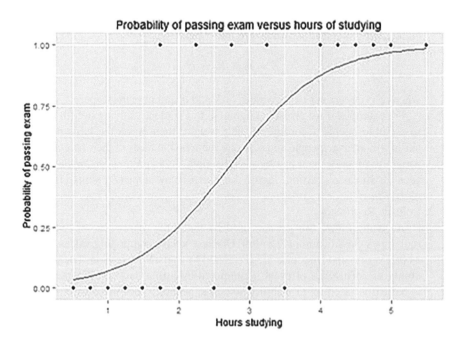

FIGURE 1.7   Logistic regression.

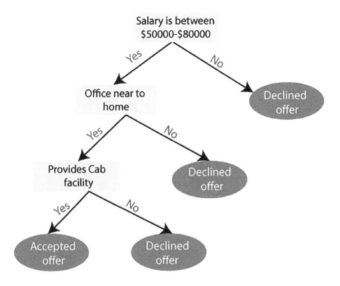

FIGURE 1.8 Decision tree.

example, if an employee is offered a salary between $50000 and $8000 and if his office is near to his home and if office provides cab facility, then the probability of that employee for taking offer letter is more, whereas if the salary is not in that range, he would have not accepted the offer; moreover, if his office was also far from his home, he would have declined the offer and if cab was not provided, still he would have declined offer (Figure 1.8).

4. SVM (Support Vector Machine)

It divides two items on the basis of their best line or decision boundary called hyperplane. In *n*-dimensional space, there can be several lines/decision boundaries to separate the groups, but we need to find the best decision boundary to help define the data points. The hyperplane of SVM refers to the best boundary (Figure 1.9).

5. Naive Bayes

A method of classification based on Bayes' theorem is called naive Bayes. This technique assumes that a particular feature in a class is not related to another. For calculating posterior probability, we use Bayes' theorem. It is given below in the form of equation:

$$T(m|n) = \frac{P(n \mid m)P(m)}{P(n)}$$

Here, $P(n|m)$ = Posterior probability
$P(m)$ = Prior probability of class
$P(n|m)$ = Likelihood which is probability of predictor
$P(n)$ = Prior probability of predictor.

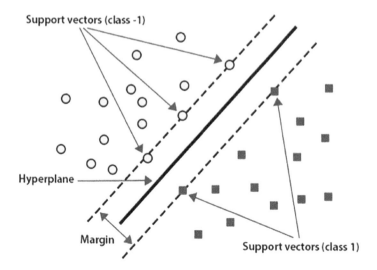

FIGURE 1.9   Support vector machine.

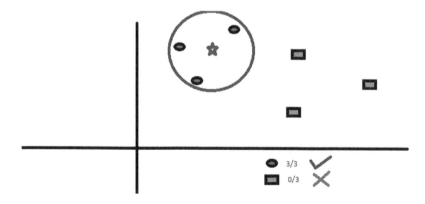

FIGURE 1.10   *k*-Nearest neighbours.

6. *k*NN (*k*-Nearest Neighbours)

It is a classification problem using classification and regression problems. *k*-Nearest neighbours algorithm involves finding the distance from the data points, and for that, we use Euclidean, Manhattan and Hamming distances. For the sake of convenience, we take an odd value of *k* such as 3 or 5 to distinguish between two different types of items (Figure 1.10).

7. *k*-Means

For solving clustering problem, we use this type of unsupervised algorithm. With the help of certain number of clusters, we can classify the data set using this technique assuming *k* number of clusters; therefore, its name became *k*-means algorithm. Figure 1.11 below depicts three prominent clusters where each cluster is shown by same coloured data points.

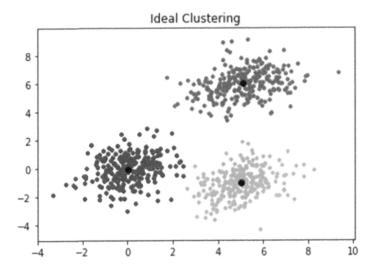

FIGURE 1.11 *k*-Means clustering.

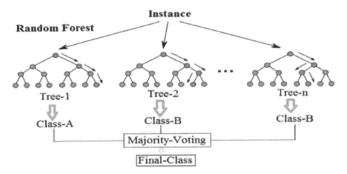

FIGURE 1.12 Random forest.

8. Random Forest

When we talk about ensembling, then random forest is the most widely used algorithm in supervised machine learning. A collection of decision trees is called a random forest. Classification is given in tree for classifying new object, and we say tree "votes" for that class. These have much more accuracy with respect to decision trees, but lower than gradient boosted trees (Figure 1.12).

1.4 RELATED WORK

There are several works done by several researchers in the field of collaborative filtering-based recommender system. Most of the work based on movie recommendation is based on the concept of personalisation, which suggests movies to users on the basis of their interest and likings.

A *k*-means clustering-based hybrid recommender system was proposed by Katarya Rahul [13] and was applied to the MovieLens data set with optimisation technique of bio-inspired artificial bee colony.

Ponnam et al. [14] suggested a collective filtering technique based on an item that examines the user's item rating matrix and determines the relationship between different objects in order to calculate the user's recommendations.

A content-based movie recommender framework was proposed by Bagher Rahimpour Cami et al. [15] capturing user choices in temporary mode in user modelling and predicting favourite movies.

Reddy et al. [16] used a genre correlation technique by using the method of content-based filtering.

A weighted hybridisation-based hybrid recommender system was proposed by Hong-Quan Do et al. [17], which didn't use fixed weight and aimed to provide a simple way to dynamically weight the combination of Collaborative Filtering and Content Based Filtering.

An effective GCN (graph convolutional network) algorithm was suggested by Rex Ying et al. [18]. The developed algorithm was effective for data that combine graph convolutions and efficient random walks to produce embeddings incorporations.

A method for tweets recommendation was proposed by Arisara Pornwattanavichai et al. [19], which was based on hybrid recommendation with LDA for unsupervised topic modelling and GMF for supervised learning.

For gaining feedback on movies and movie genres in Rohan Nayak et al. [20] hybrid's framework, and based on their responses, the user will be classified and given a collection of recommendations.

Collaborative filtering, as previously discussed, is a well-known technique for making powerful recommendations based on ratings results. In order to enhance the technique's ability and achieve results by *k*-means clustering algorithm in movie recommendation framework, we continue our research.

1.5 MACHINE LEARNING TECHNIQUES & METHODOLOGY USED FOR RELIABILITY ASSESSMENT

The entire machine learning process is divided into several tasks. The first and foremost task is data set identification, and we have chosen MovieLens data set for our experimentation. From the well-known GroupLens Research Project at the University of Minnesota, we took MovieLens data [21]. Our goal with using this data set is to generate recommendations of movies to users on the basis of their interest and likings. This data set comprises 264505 ratings (1–5 scale) from 862 users on 2500 movies, and age, occupation, zip code, gender, etc., act as important demographic features taken from user data set. Next, data preprocessing is done to remove any sort of noise from the data set.

For our experimentation work, we are splitting the data set into two parts by 80:20, where the training part (80%) is used to train our model and then 20% is used

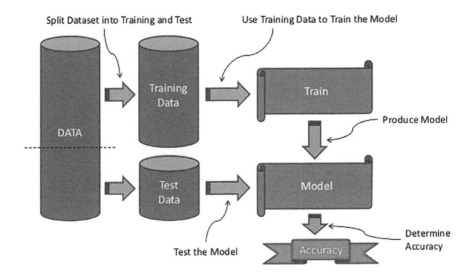

FIGURE 1.13   Machine learning process.

TABLE 1.1
Details of MovieLens Data set

Data set Name	Number of Unique Data
Movies.CSV	2500 Movies
Ratings.CSV	264505 Ratings
Users.CSV	862 Users

for testing. Finally, we also evaluate our model by calculating RMSE and MAE of our proposed model (Figure 1.13).

1.5.1   DATA SET

We have taken MovieLens data set for our experimental work. This data set has been taken from (http://www.movieLens.org) for evaluating our proposed recommender system. Our experiments are performed on Google Colab where Google provides with the support of hardware on cloud to do our machine learning task. Here ratings by users are given on a scale from 1 to 5. Our data set is comprised of those users who have given at least 20 ratings. Our data set comprises 1,000,209 ratings given by users for different movies (Table 1.1).

1.5.2   COLLABORATIVE FILTERING TECHNIQUE

This approach is based on a user's suggestion of an object based on reactions from similar users. This works by selecting a smaller collection of users from a wide

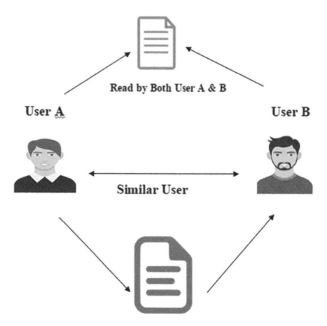

Read by Both User A & B

User A   User B

Similar User

Read by User A, Recommended to User B

FIGURE 1.14   Collaborative filtering technique.

community of individuals with tastes close to a single user. In this, the main recommendation principle is that other users offer ratings to a specific object (Figure 1.14).
    Measuring user similarity in collaborative filtering technique:

i. Pearson Correlation:

$$\sin(a,b) = \frac{\Sigma_{P\in P}(ra.p - ra)(rb.p - rb)}{\sqrt{\Sigma_{p\in P}(ra.p - ra)^2}\sqrt{\Sigma_{p\in P}(rb.p - rb)^2}} \tag{1.4}$$

where a and b are users, while $r_{a.p}$ is rating and P is set of items read by both users.

ii. **Cosine Similarity Measure**: It is measured by the angle between the vectors

$$\sin(\vec{p},\vec{q}) = \frac{\vec{p}.\vec{q}}{|\vec{p}| * |\vec{q}|} \tag{1.5}$$

U represents users having rated both items p and q.

1.6   EXPERIMENTAL SET-UP

The idea behind recommending movies to users based on item-item collaborative filtering comprises the steps discussed below:

Step 1. Create an adjusted rating for all movies by users. This adjusted rating is calculated by subtracting the movie's average rating from all users (for movie j) from each rating for that movie.

Step 2. Calculate similarity scores between all movies based on their adjusted movie ratings from each user (use cosine similarity). For recommendation purpose, we will only consider top similar movies to a target movie (top n nearest neighbours).

Step 3. For recommending a movie to a target user, we will score each movie, using the top n nearest neighbours for that movie. The score is basically a weighted rating based on the target user's rating for all movies they have rated and the similarity scores as the weight. Once we score all the movies, pick the top scoring movies from this scoring as recommendations.

The adjusted rating is nothing but the average rating for the movie from all users (uj) subtracted from all of the individual movie ratings (ru, j):

$$Ru, j = ru, j - uj$$

This adjusted rating is now comparable across all movies. This adjusted score basically compares the variation of ratings by a user from the movie's mean rating (Figures 1.15 and 1.16).

Now we create similarity score for each movie with every other movie; for this, we use the concept of cosine similarity (Table 1.2).

For creating recommendation to the target user, we find a score for each movie in the data set and movies with the highest score will be recommended to the user.

Steps involved in scoring are as follows:

1. Get the list of movies the target user has rated (seen movies). These seen movies will be used to create the score for all other movies (unseen movies) based on how the unseen movies are similar to these seen movies. These

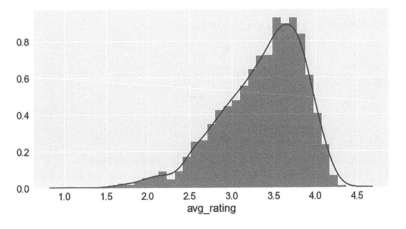

FIGURE 1.15 Potting average ratings across all users.

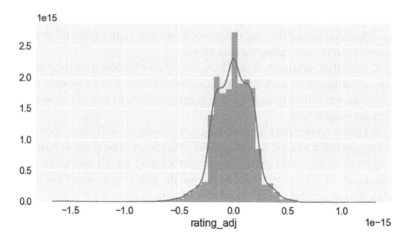

FIGURE 1.16 Potting average-adjusted ratings across all users.

seen movies tell us about the taste of the target user. If they have rated some movies high, we will try to find similar unseen movies to these high rated seen movies and recommend them to the user and vice versa for low rated movies.

2. For all the unseen movies in the data set, get the similarity scores between them and the seen movies. Here we can use all the seen movies or the top N neighbours out of the seen movies to get the similarity scores. We will use $N = 30$ for our calculation. In case the number of seen movies is less than 30, we will use all the seen movies.

3. Using the similarity scores between each of the unseen movies and the seen movies, calculate a score for the unseen movies. The formula for the score is given below.

4. Once we get the score, sort the unseen movies based on the score and recommend the top n movies for the user.

We use the following formula to calculate score:

$$S_{u,i} = m_u + \frac{\sum_{j} \cos(i,j).(r_{uj} - m_j)}{\sum_{j} \cos(i,j)}$$

where

S is the score for the unseen movie i

m_u is the average rating for all seen movies by the target user U

$\cos(i,j)$ is the cosine similarity (based on adjusted rating) between the unseen movie i and the seen movie j

r_{uj} is the rating of the seen movie j by the target user U

m_j is the average rating from all users for the seen movie j

$r_{uj} - m_j$ is the same as the adjusted rating calculated above.

TABLE 1.2
Similarity Score of Movie vs Movie

Movield	1	2	3	4	5	6	7	8	9
1	1.000000	0.213859	0.141760	−0.008966	0.097387	0.142986	0.098391	−0.002693	0.249048
2	0.213859	1.000000	0.218855	0.038701	0.125331	0.088945	0.154515	0.087974	0.231964
3	0.141760	0.218855	1.000000	0.056912	0.194855	0.067841	0.215001	0.084497	0.238945
4	−0.008966	0.038701	0.056912	1.000000	0.130774	0.014619	0.165135	0.008468	0.002328
5	0.097387	0.125331	0.194855	0.130774	1.000000	0.014217	0.135021	0.033035	0.070476

TABLE 1.3

Recommendations for User 76630

	Movield	Title	Genres	Score
0	2906	Random Hearts (1999)	Drama|Romance	3.086117
1	1099	Christmas Carol, A (1938)	Children|Drama|Fantasy	3.060448
2	828	Adventures of Pinocchio, The (1996)	Adventure|Children	3.040377
3	611	Hellraiser, Bloodline (1996)	Action|Horror|Sci-Fi	3.018605
4	1015	Homeward Bound: The Incredible Journey (1993)	Adventure|Children|Drama	3.005596
5	334	Vanya on 42nd Street (1994)	Drama	2.985227
6	3684	Fabulous Baker Boys, The (1989)	Drama|Romance	2.978881
7	1014	Pollyanna (1960)	Children|Comedy|Drama	2.976269
8	1218	Killer, The (Die xue shuang xiong) (1989)	Action|Crime|Drama|Thriller	2.974656
9	2859	Stop Making Sense (1984)	Documentary|Musical	2.970456

TABLE 1.4

Splitting Data set into Training and Testing

Number of Users, Ratings and Movies	Training Data	Testing Data
Number of unique users in RATINGS data	681	181
Number of ratings in RATINGS data	209235	55270
Number of movies	2500	2496

Both the test and training data sets show similar distribution for the number of users per movie and average rating per movie (Tables 1.3 and 1.4). This shows that the test and training data sets are not that different and should be good enough for our evaluation. There is difference in the distribution of the average movie rating per user in test and training data sets, but these should be OK as we will use adjusted movie ratings for our recommendations (Figures 1.17–1.20).

1.6.1   Test Data Set – QUERY vs PROBE

Even from the given test data set, while trying to get the prediction for one user, we will only keep some movie ratings away from the model (QUERY movies), while we will pass on the remaining movies from that user to the model to be used as history (PROBE movies).

This division can be done randomly or on a temporal basis. We will do this based on time (temporal) – keep most recent ratings from a user as query and the older ones as probe. We can do this based on the timestamps available in the ratings data set.

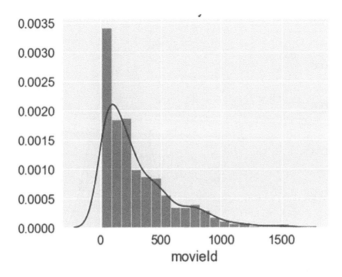

FIGURE 1.17   Movies rated by user in training data set.

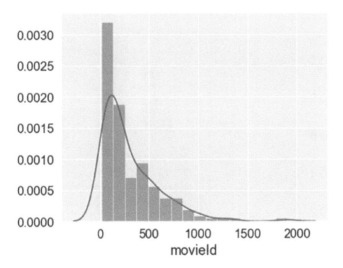

FIGURE 1.18   Movies rated by user in testing data set.

Algorithm 1. User-User Collaborative Filtering

The complete algorithm for user-user CBF will be explained in the following defined function. The steps for this algorithm are the following:

1. Create adjusted user movie rating.
2. Create similarity score for each user with every other user.
3. Create recommendation for the target user based on the similarity score.

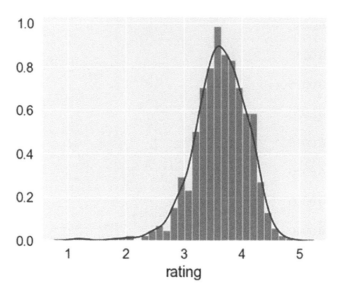

FIGURE 1.19   Average movie rating in training data set.

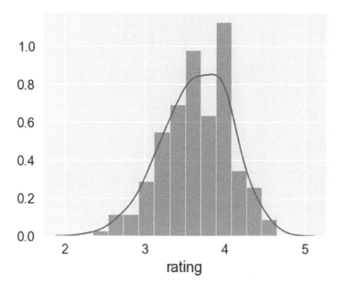

FIGURE 1.20   Average movie rating in testing data set.

Algorithm 2: Item-Item Collaborative Filtering

The steps for the item-item CBF will be as follows:

1. Create adjusted rating for every movie.
2. Get similarity scores between every movie.

3. Rank each movie for a given target user based on a score created using similarity scores between the movie and the top neighbours of the movies (which target user has rated).

1.7   RESULTS EVALUATION

1.7.1   EVALUATE THE RECOMMENDATION FROM BOTH ALGORITHMS – RMSE AND MAE

In our test query ratings data set, we loop through all the users and get the recommendation from both the algorithms. We will then use the predicted ratings for their movies and compare them with their actual rated movies to calculate the RMSE (root-mean-square error) and MAE (mean absolute error) metrics. The algorithm with the least RMSE or MAE will be considered better performing.

The graph below depicts a comparison between item-item CBF and user-user CBF with the number of neighbours with respect to RMSE (Figure 1.21).

The graph below depicts a comparison between item-item CBF and user-user CBF with the number of neighbours with respect to MAE (Figure 1.22).

From the above graph, it's pretty clear that the user-user algorithm gives much better prediction than the item-item algorithm. It also looks like that the neighbourhood size of ~20 is good enough in our case for user-user algorithm.

We are not choosing the neighbourhood size of 5 as it basically gives out very less number of recommendations and is not good enough.

The table below depicts the RMSE and MAE comparison table the two algorithms item-item CBF and user-user CBF (Table 1.5).

The graph given below depicts comparison between RMSE and MAE with respect to the two algorithms item-item CBF and user-user CBF (Figure 1.23).

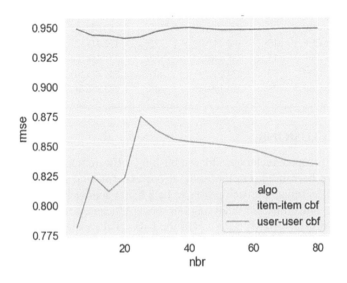

FIGURE 1.21   RMSE plot for algorithms.

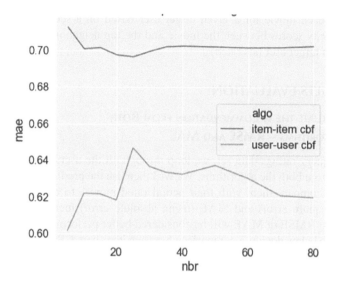

FIGURE 1.22   MAE plot for algorithms.

TABLE 1.5
RMSE and MAE Comparison Table

	Algorithm	NBR	Error_sq	Movield	Error_abs	RMSE	MAE
0	Item-item CBF	5	489.039404	543	386.951160	0.949013	0.712617
1	Item-item CBF	10	483.644746	543	380.510503	0.943764	0.700756
2	Item-item CBF	15	483.149094	543	380.910301	0.943280	0.701492
3	Item-item CBF	20	480.815570	543	378.782391	0.940999	0.697573
4	Item-item CBF	25	482.042223	543	378.205743	0.942199	0.696511
12	User-user CBF	5	81.272629	133	80.028456	0.781711	0.601718
13	User-user CBF	10	188.447988	277	172.318323	0.824814	0.622088
14	User-user CBF	15	235.440337	357	221.988758	0.812094	0.621817
15	User-user CBF	20	274.206791	404	249.881260	0.823851	0.618518
16	User-user CBF	25	340.749089	445	287.846705	0.875059	0.646847

1.8   CONCLUSIONS

In the present chapter, techniques for establishing software reliability using machine learning have been used. On the basis of our experimental results, it is revealed that machine learning approach proves to be a better approach for predicting accurate software reliability. For analysing our model efficiency, we use the concept of RMSE, NRMSE and MAE criteria. On the basis of the experiment conducted on the well-known MovieLens data set, the ML approach gives better results and it is revealed that our technique provides more accurate results. The results obtained from our experimentation work reveals that the ML-based approach decreases testing cost by estimating the reliability of software and is much more feasible.

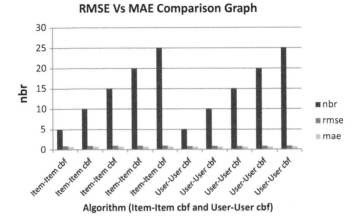

FIGURE 1.23   RMSE and MAE comparison graph for item-item CBF and user-user CBF w.r.t NBR.

REFERENCES

[1] Schick, G.J., Wolverton, R.W., 1978. An analysis of competing software reliability models, *IEEE Trans. Softw. Eng.* SE-4(2), 104–120.

[2] Yamada, S., 2014. *Software Reliability Modeling: Fundamentals and Applications*, Springer, Heidelberg. https://doi.org/10.1007/978-4-431-54565-1.

[3] Almering, V., van Genuchten, M., Cloudt, G., 2007. Using software reliability growth models in practice, *IEEE Comput. Soc.* 24, 82–88.

[4] Quadri, S.M., Ahmad, N., Farooq, S.U., 2011. Software reliability growth modeling with generalized exponential testing effort and optimal software release policy, *Glob. J. Comput. Sci. Technol.* 11(2), 27–42.

[5] Freedman, D.A., 2009. *Statistical Models: Theory and Practice*, Cambridge University Press.

[6] Edwards, J.R., 2007. Polynomial regression and response surface methodology. *Perspect. Organ. Fit*, pp. 361–372.

[7] Smola, A.J. and Schölkopf, B., 2004. A tutorial on support vector regression. *Stat. Comput.*, 14(3), 199–222.

[8] Lewis, R.J., 2000. An introduction to classification and regression tree (CART) analysis. In *Annual Meeting of the Society for Academic Emergency Medicine in San Francisco, California*, Vol. 14. Department of Emergency Medicine Harbor-UCLA Medical Center Torrance, California.

[9] Liaw, A. and Wiener, M., 2002. Classification and regression by random Forest, *R News*, 2(3), 18–22.

[10] Minh, L.Q., Duong, P.L.T. and Lee, M., 2018. Global sensitivity analysis and uncertainty quantification of crude distillation unit using surrogate model based on Gaussian process regression, *Ind. Eng. Chem. Res.*, 57(14), 5035–5044.

[11] Kabir, H.D., Khosravi, A., Hosen, M.A. and Nahavandi, S., 2018. Neural network-based uncertainty quantification: A survey of methodologies and applications, *IEEE Access*, 6, 36218–36234.

[12] Qiu, X., Zhang, L., Ren, Y., Suganthan, P.N. and Amaratunga, G., 2014. Ensemble deep learning for regression and time series forecasting. In *2014 IEEE Symposium on Computational Intelligence in Ensemble Learning (CIEL)* (pp. 1–6). IEEE.

[13] Katarya, R., 2018. Movie recommender system with metaheuristic artificial bee, *Neural Computing and Applications* 30(6), 1983–1990.

[14] Ponnam, L. T., Deepak Punyasamudram, S., Nallagulla, S. N., Yellamati, S., 2016. Movie recommender system using item based collaborative filtering technique. In *2016 International Conference on Emerging Trends in Engineering, Technology and Science (ICETETS)*. IEEE, Pudukkottai, India, pp. 1–5.

[15] Cami, B. R., Hassanpour, H., Mashayekhi, H., 2017. A content-based movie recommender system based on temporal user preferences. In *2017 3rd Iranian Conference on Intelligent Systems and Signal Processing (ICSPIS)*. IEEE, Shahrud, pp. 121–125.

[16] Reddy, S., Nalluri, S., Kunisetti, S., Ashok, S., Venkatesh, B., 2019. Content-based movie recommendation system using genre correlation. In *Smart Intelligent Computing and Applications. Smart Innovation, Systems and Technologies*. Springer, Singapore, pp. 391–397.

[17] Do, H.-Q., Le, T.-H., Yoon, B., Dynamic weighted hybrid recommender systems. In *2020 22nd International Conference on Advanced Communication Technology (ICACT)*. pp. 644–650.

[18] Ying, R., He, R., Chen, K., Eksombatchai, P., Hamilton, W. L., Leskovec, J., 2018. Graph convolutional neural networks FORWEB-scale recommender systems. In *Proceedings of the 24th ACM SIGKDD International Conference on Knowledge Discovery & Data Mining. KDD '18*. Association for Computing Machinery, London, UK, pp. 974–983.

[19] Pornwattanavichai, A., Sakolnagara, P. B. N., Jirachanchaisiri, P., Kitsupapaisan, J., Maneeroj, S. Enhanced tweet hybrid recommender system using unsupervised topic modeling and matrix factorization-based neural network. In *Supervised and Unsupervised Learning for Data Science*. Springer International Publishing, pp. 121–143.

[20] Nayak, R., Mirajkar, A., Rokade, J., Wadhwa, G., 2018. Hybrid recommendation system for movies, *Int. Res. J. Eng. Tech.* 05(03), 4.

[21] Cabannes, V., Rudi, A., Bach, F., 2021. Disambiguation of weak supervision with exponential convergence rates. *Proceedings of the 38th International Conference on Machine Learning*, ICML, arXiv:2102.02789.

2 Significance of Machine Learning and Deep Learning in Development of Artificial Intelligence

D. Akila
Saveetha College of Liberal Arts and Sciences,
SIMATS Deemed to be University

S. Jeyalaksshmi, D. Padmapriya,
Devipriya and Piramu Prithika
Vels Institute of Science, Technology and Advanced Studies

V.R. Elangovan
Agurchand Manmull Jain College

CONTENTS

2.1 INTRODUCTION

Machine learning is a subtype of AI that allows a machine to study without explicitly programming concepts or facts. It starts with personal observations in order to anticipate data features and trends and give superior results and judgements in the future. Deep learning is a set of machine learning techniques that utilise a large number of nonlinear transformations to represent high-level abstractions in data [1]. ML is an artificial intelligence (AI) discipline. DL is not a new concept, and it has been criticised as a rebranding of neural networks. Recent research in the use of DL for Mobile Device Management, however, has proved promising, particularly in the field of visual data mining [2].

Deep learning has been one of the most significant technological advancements in the field of artificial intelligence during the last 10 years. Deep learning, in contrast to shallow learning, often necessitates a high number of neuronal layers. Deep learning models are superficial learning methods in computer vision, speech recognition, automated machine translation and financials [3–5]. Deep learning's advantages in other domains have led researchers to apply it to intrusion detection. Traditionally, the entire data set is used to train a single deep learning model. Pre-trained deep learning models are constructed models that help people learn about algorithms or experiment with current frameworks for better outcomes without directly building them [6–8].

Deep learning methods employ a sequential layer architecture to automatically extract features from a data set. The introduction to the sequential layer structure of nonlinear transformation functions is the basis of deep learning techniques. The complexity of producing nonlinear transformations increases as the number of layers increases. Deep learning methods employ abstract representations at several layers to understand the hidden abstract features of the data collected from the final layer. This results in the input being routed through a high-level nonlinear function to provide abstract properties for the last output layer. The basis of SVM techniques is statistical and convex learning, which are founded on the concept of structural risk reduction. SVM was invented by Vapnik as a solution to a number of problems. It may be used for learning, pattern recognition, regression, classification and analysis, among other things [9]. Deep learning is an emerging approach of machine learning that can handle enormous data sets and actual words. The basic conception of deep learning is based on neural networks. The input and output levels are concealed over four layers (also known as nodes). In recent times, deep learning has become popular among academics due to its capacity to deal with enormous volumes of data and difficult issues such as voice, video, picture and audio. DL can also handle categorisation problems such as time series and computer vision [10].

Big data is becoming more accessible in many aspects of production and operations. Data, in and of itself, have value in allowing a competitive data-driven economy, which is at the heart of the Internet of things and Industry 4.0. The increased data availability allows for improved decision-making and strategy formulation, as well as the introduction of the next generation of creative and disruptive technologies [11]. Computer learning is a discipline in which a computer recognises numerous components or parts of the data supplied as input and then the system produces output predictions. Machine learning is an area of research that combines artificial intelligence with statistics to allow computers, depending on the input, to predict and process output. Machine learning is now divided into three kinds: supervised, unattended and individually characterised. The input is supplied in the form of numerous examples in supervised machine learning, and the machine aims to get the desired output by evaluating, measuring and calculating various parameters in the input. The type of data collection used for input, biasing and labelling, and the algorithm employed to interfere with the input are all factors that impact the prediction's accuracy [12].

Deep learning (DL) is both a new academic buzzword and a machine learning discipline (ML). ML is connected with some of artificial intelligence's (AI) basic ideas and focuses on addressing real-world issues with neural networks to replicate our own decision-making processes. The words "deep" and "learning" are combined. Deep means that something goes deep below the surface; it usually refers to the number of layers, and learning refers to the acquisition by study and practice of information or abilities. DL is a form of learning data representation based on more advanced (hierarchical) knowledge. We may think of deep learning as learning hierarchical representations. There are various types of learning, including supervised, unsupervised, and partially supervised learning. DL, also known as hierarchical learning or deep structured learning [13], is a subset of ML. Deep learning has shown a lot of promise as a replacement for handcrafted features in computer vision jobs in recent years [14].

Deep learning, a branch of artificial intelligence (AI), is currently a popular and widely used method that has been used in fields such as biology, medicine, computer vision and speech recognition. Deep learning is a relatively new AI technique that provides a strong framework for supervised learning. Even with large data sets, it can rapidly and efficiently convert an input vector to an output vector. Deep learning architectures such as the convolutional neural network (CNN) and deep belief network (DBN) are available. Deep learning can extract more comprehensive information than traditional machine learning algorithms [15]. This study also discusses the key distinctions between deep learning and machine learning in terms of the importance of developing an artificial intelligence-based picture classification and recognition framework for large data. Deep learning and machine learning classifications are coupled to improve picture classification performance on huge data.

2.2   RELATED WORKS

Pournami S. Chandran et al. [16] presented a first-class approach to locating a kid missing from the photographs of a vast number of children with facial recognition a profound learning technique. The entire public may upload photographs of dubious youngsters, as well as landmarks and remarks, to a shared site. The photograph will be promptly matched to the repository's photographs of the missing kid. The photograph of the missing child is classified, and a database of missing children collects the best match. The missing child in the missing photograph database has a deep learning model designed by the public to accurately identify the missing child with a face photograph. A particularly successful deep learning approach for image-based applications, the CNN, is used to identify face. Visual descriptors are removed from pictures with the VGG-Face deep architecture, a pre-trained CNN model. Unlike traditional deep education applications, our technique simply utilises a convolution network as a high-level functional extractor, which handles child detection using a trained SVM classifier. The most efficient CNN model for face recognition, VGG-Face, has been picked and appropriately trained and has resulted in a deep learning model, which is insensitive to noisy, lightning, contrast, occlusion, picture posture and childhood. The kid identification system's classification performance is 99.41%. It was put to test on 43 kids.

Dong Yu-nan et al. [17] in their paper compared deep learning to traditional machine learning methods, then described the deep learning development process, investigated and analysed deep learning network structures such as DBN, CNN and RNNs, elaborated on the use of deep learning in image recognition and classification, and proposed deep learning. The problems of using recognition and classification are addressed, as well as the answers to those problems. Finally, the present status of research in deep learning for image identification and classification is presented, as well as future prospects.

In the study of Sufri et al. [18], two studies were carried out on two sorts of pictures from the banknote: different areas and guidance captured in a controlled environment using a smartphone camera and separate regions and orientations recorded in a controlled environment using a smartphone camera. Machine learning modules have been taught to recognise each banknote class by removing feature values RB, RG and GB from banknote pictures using different methods such as k-nearest neighbours (kNN), decision tree classifier (DTC), support vector machine (SVM) and Bayesian classifier (BC). AlexNet is a prequalified model of the CNN, the most common structure for image processing in deep learning NN.

Yanyan Dong et al.'s [19] approach focused on the stage of extracting features from a retinal picture. To begin, the fundus pictures are pre-processed using the maximum entropy approach. Then, using a Caffe-based deep learning network, we extract more differentiating features from fundus images automatically. A range of classification techniques are finally utilised to automatically identify derived characteristics. Instead of deep learning characteristics and features derived from the retinal vascular, SVM (support vector machine) and softmax are utilised for cataract classification. Cataract pictures are finally categorised as normal, moderate or severe. When compared to classification results, the feature retrieved through deep learning and categorised using softmax has higher accuracy. The findings show that our deep learning research is both successful and useful.

Jiang Huixian et al. [20] reported a comparison of 50 plant sheet data sets with the KNN classification bases, the Kohonen network and the SVM based on a self-organising approach for the mapping of features. The leaves were compared to seven different plants at the same time, and the ginkgo leaves were found to be simpler to recognise. A good recognition effect has been achieved for leaf pictures with complicated backgrounds. Image samples from the test set are entered to retrieve reconstruction faults in the learning model. The deep learning model with the lowest error set is decided by the test set's class label. This technique offers the fastest possible identification time and the highest correct identification rate based on the data.

Busra Rumeysa Mete et al. [21] using deep CNN and data augmentation proposed a classification method for floral photographs. Deep CNN techniques have recently emerged as the most advanced solution for dealing with such issues. However, gaining improved performance for flower categorisation is hampered by a shortage of labelled data.

Hossam M. Zawbaa et al. [22] aimed to build an efficient and effective classification approach based on the RF algorithm. Different traits have been identified based on the form, colour and scale invariant characteristics for classifying three fruits: apples, strawberries and oranges. An image processing pretreatment phase is presented to prepare the fruit pictures by decreasing their colour index. The visual features of the fruit will then be found. Finally, random forest (RF), a freshly created machine learning method, is used in the fruit categorisation process. The photographs were captured using a standard digital camera, and all changes were carried out in a MATLAB® environment. Trials were performed and reviewed using 178 fruit photographs in a series of experiments. From the standpoint of accuracy, the RF technique demonstrates that the methodology can also be used to improve other famous algorithms such as kNN and SVM methods. In addition, the system is highly accurate in the recognition of the fruit name automatically.

Mohit Sewak et al. [23] examined one of the deep learning architectures, the deep neural network (DNN), and compared it with the conventional RF malware classification learning technique. We tested traditional RF and DNN performance with two-, four- and seven-layer architectures and four feature sets. The classical RF exceeds DNN irrespective of the feature inputs.

Obesity is the major cause of stroke and death in many nations [24]. Data pre-processing was used to enhance the image quality of CT scans by stretching picture quality for improving image results and reducing noise. It also utilised machine learning algorithms for classifying images of patients into two categories of stroke disease: ischaemic stroke and haemorrhagic stroke. The eight machine learning algorithms employed in the trial to identify stroke illness were kNN, naive Bayes, logistic regression, decision tree, random forest, multilayer perceptron (MLP-NN), deep learning and SVM. Random forest, according to the results of the research, delivered the highest level of accuracy.

Kyu Beom Lee et al. [25] developed an object detection and tracking system (ODTS) with the well-known deep learning network faster region convolutional neural network (Faster R-CNN) and object detection and conventional object tracking. They used the developed system to automatically detect, and monitor unexpected events in tunnel CCTVs.

Kavitha et al. [26] presented a CNN architecture for separating various plant pictures from collected sequences. In order to remove the features of the pictures in the resultant data set, CNN architecture is employed after the pre-processing procedures, which may include removing bleakness or adding a lighting shift. Create a data set, train CNN, validate, test CNN, and predict and categorise the photograph are all phases in the process of image classification using CNN. In one case, the whole classification report is anticipated with a precision of 43.98% by using Keras software with the Theano and the TensorFlow backend implementations.

Qing Li et al. [27] developed a tailored CNN with a shallow convolution layer to categorise pulmonary photographic patches with interstitial lung disease (ILD). Although in recent years a host of feature descriptors have been suggested, they can be rather complicated and domain specific. On the other hand, our proprietary CNN framework can learn the intrinsic picture functions from lung image parts, which are most suitable for classification, automatically and effectively. Different tasks in the categorisation of medical images or texture may be performed using the same framework [28–30].

2.3 PROPOSED SYSTEM

In recent years, the quantity of data has quickly increased due to the increased usage of social media and Internet of things (IoT) appliances that require sensors, networks and communications technology. The documents are available in both structured and unstructured formats and, without suitable techniques and tools, are difficult to handle. Researchers have devised a number of techniques and technologies aimed at coping with large amounts of data. For example, the Apache Software Foundation built Hadoop and Spark in a parallel processing architecture to address massive amounts of data. Other solutions being developed to address complex data issues are Google Dremel and S4. To enhance decision-making, all of these technologies are utilised to collect, analyse and interpret huge amounts of data [29,30]. New thoughts and approaches for dealing with diverse big data challenges are offered on a regular basis. Deep learning is a novel method to machine learning, which can handle huge data and real-world scenarios. Deep learning's underlying notion is derived from neural networks. Deep learning and machine learning have recently gained interest among researchers for addressing massive data and associated challenges, and they are used in various applications such as acoustic modelling, adaptive testing, automotive industry, big data, biological image classification, data flow graphs, deep vision system, document analysis and recognition, healthcare, human activity recognition, image recognition and classification, medical applications, mobile multimedia, object detection, parking system, plant classification, semantic image segmentation, stock market analysis and structural health monitoring.

This section illustrates some of the machine learning and deep learning models used on the images collected from the big data. The images are taken from the big data for further development of an artificial intelligence method for image classification and recognition for further process. Figure 2.1 below shows the proposed framework of this paper.

The input images are taken from the big data. And the images are further processed by pre-processing, feature extraction and classification algorithms.

2.3.1 IMAGE PRE-PROCESSING

Because of the disparity in the quality of photographs, it is essential to perform picture preparation. Picture enhancement is an important step in the fundus image pre-processing. We utilise histogram equalisation before; however, this approach loses a lot of visual information and does not accentuate the blood vessels in the backdrop image. After several testing runs, we choose the biggest entropy transformation. In order to identify the correct grey level categoriser, threshold of the image and transform feature will execute the local grey level transformation based on the concept of maximum entropy.

Locate the optimum dividing point to acquire the grey image edge, and then achieve the threshold value on each side and enhance the spatial nonlinear deformation function (2.1 and 2.2). Not only does this approach enhance the image quality, but it also saves time and preserves as much of the information from the original fundus picture as possible.

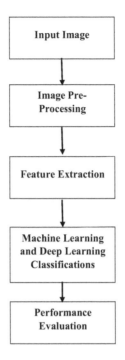

FIGURE 2.1   The block diagram of the proposed algorithm.

The following is the formula:

$$H(x) = \sum_{i=1}^{k} p(x = xi) \log \frac{1}{p(x = xi)} \qquad (2.1)$$

$$0 \le H(x) \le \log|x| \qquad (2.2)$$

2.3.2   FEATURE EXTRACTION

This stage is aimed towards extracting the characteristics or properties of an image. The degree of extraction influences the categorisation accuracy.

In consequence, the research suggested examines two ways of obtaining picture characteristics: form and colour properties and the scale invariant feature transform (SIFT) [22].

Colour is regarded as an important characteristic for image representation as it is invariant in the translation, scaling and rotation of images. As a result, for each fruit image in the collection, the first feature extraction approach utilised its colour and form properties. Colour variation, colour mean, colour kurtosis and colour skewness are all colour moments used to characterise the photographs. The terms eccentricity, centroid and Euler number are used to describe the form characteristics. Eccentricity is computed by dividing the main axis distance by the minor axis distance. It's calculated using either the major axis technique or the minimum bounding rectangle technique. The centroid of the picture in respect of the shape is specified by the shape centroid. The connection between the number of connecting parts and the number of holes on an image form is established by the image Euler number. To get the Euler number, subtract the number of form holes from the number of contiguous sections.

The SIFT technique is used in the second technique to build the feature vector. It is a way to extract visual characteristics that are not sensitive to image rotation, scaling and translation, as well as to sophisticated projection and lighting changes.

SIFT is broken down into four main phases: All examples of extreme scale space detection are key point location, orientation assignment and descriptor key point. The extreme discovery stage in scale space identifies possible locations of interest utilising the gauze-like function. A model is designed for the position and scale of each location at the key point location. The key points selected are based on their stability criterion. Orientations are assigned to each key point location in the orientation assignment stage based on the local image gradient directions. The algorithm is mentioned below.

Algorithm 1: SIFT Feature Extraction Algorithm

Step 1: Using equations 2.1–2.3, construct the picture Gaussian pyramid $L(x, y, \sigma)$.

$$G(x, y, \sigma) = \frac{1}{2\pi\sigma^2} \exp^{\frac{-(x^2 - y^2)}{2\sigma^2}} \tag{2.3}$$

$$L(x, y, \sigma) = G(x, y, \sigma) * I(x, y) \tag{2.4}$$

$$D(x, y, \sigma) = L(x, y, k\sigma) - L(x, y, \sigma) \tag{2.5}$$

where σ is the scaling parameter and $G(x, y)$ is the Gaussian distribution.
$I(x, y)$ is a smoothing filter, while $L(x, y)$ is a Gaussian filter.
$D(x, y)$ is the Gaussian difference (DoG).
Step 2: Determine the Hessian matrix.
Step 3: Then, as described in equation 2.6, compute the determinant of the Hessian matrix and delete the weak key points.

$$\text{Det}(H) = \text{Imm}(x, \sigma)\text{Inn}(x, \sigma) - \left(\text{Imn}(x, \sigma)\right)^2. \tag{2.6}$$

Step 4: As in equations 2.7 and 2.8, calculate the gradient magnitude and direction.

$$\text{Mag}(x, y) = \left(\left(I(x + 1, y) - I(x - 1, y)\right)^2 + \left(I(x, y + 1) - I(x, y - 1)\right)^2\right)^{1/2}. \tag{2.7}$$

$$\theta(x, y) = \tan^{-1}\left(\frac{I(x, y + 1) - I(x, y - 1)}{I(x + 1, y) - I(x - 1, y)}\right) \tag{2.8}$$

Step 5: As in equations 2.9 and 2.10, use the sparse coding feature based on SIFT descriptors.

$$\min \sum_{i=1}^{s} \left(xi - \sum_{j-1}^{z} a_i^{(j)} \, \varphi^{(j)} \parallel 2 + L \right) \tag{2.9}$$

$$L = Y \sum_{j=1}^{z} \left| a_i^{(j)} \right| \tag{2.10}$$

where x_i is the SIFT descriptors feature, a_j is mainly zero (sparse), φ is the sparse coding basis, and Y is the weights vector.

The operations are then carried out on image data translated for each feature according to the specified size, orientation and location to ensure these modifications are invariant. The local picture gradients in a specified quarter surrounding the recognised key point are calculated for the selected scale in the key point descriptions step. These are translated into representation to allow significant amounts of deformation of local form and changes in lighting. It follows Algorithm 1's instructions.

2.3.3 CLASSIFICATIONS

2.3.3.1 Support Vector Machine

Support vector machine (SVM) is a simple and effective supervised learning technique used in categorisation. However it is frequently utilized in categorisation process.

In SVM, the distribution of input data is not known or assumed in advance. There are two possible methods for dividing the data: linear and nonlinear methods. Furthermore, using SVM, there is no overfitting. If cross-validation is not done, overfitting may occur in artificial neural networks. Several kernel functions can also be utilised in a more readable space to isolate indivisible issues and map data. Kernel-based algorithms are highly flexible because the methods have no influence on hyperspecific factors, including the learning rate and parameters. Another explanation is that changing the kernel function is sufficient when the issue area changes.

The equations provided were the kernel functions most commonly employed, such as polynomial (1), linear (2) and Gaussian (3) kernels.

$$K(u, v) = (u \times v + 1)^d \tag{2.11}$$

$$K(u, v) = u \times v \tag{2.12}$$

$$K(u, v) = \exp\left(\frac{- \parallel u - v \parallel^2}{2\sigma^2} \right) \tag{2.13}$$

The SVM classifier proved to be the most successful in this research. We found that after testing with several kernels such as polynomial, linear and Gaussian/RBF, we achieved a maximum accuracy of 93.2% in 116 seconds on the Food-101 data set. Using the sklearn LinearSVC module with default parameter values, we were able to come up with these results.

2.3.3.2 Convolutional Neural Network

Convolutional neuron layers are frequently used by CNN. One or more 2D matrices (or channels) in image classification tasks are entered into the convolutional layer, producing numerous 2D matrices. The number of inputs and output matrices may vary. The following is the process for calculating a single output matrix:

$$Aj = f\left(\sum_{i=1}^{N} Ii * Ki, j + Bj\right) \qquad (2.14)$$

Each input matrix Ii is first twisted with a matching kernel matrix Ki, j. The total of all convoluted matrices is then computed, and each member of the resultant matrix is given a bias value Bj. Finally, each component of the previous matrix is subjected to a nonlinear activation function fi, resulting in a single output matrix Aj (Figure 2.2).

CNN's fundamental structure consists of two levels: The first is connected to the local windows of the preceding layers to extract features and each neuron in one layer. The second layer is the mapping layer of the feature. CNN is employed as it is resistant to change in picture and distortion, takes less memory, is easier to use and gives a more effective model of training. During the image processing and voice recognition, it has greater importance since it gives a unique structure through shared local weights and is almost identical to that of a biological neural network.

The overlay layer consists of a series of filters, which are separately intertwined with the picture input. All filters are set up randomly at the beginning, and then the network screens their coefficients. The output of the neurons related to the local input areas is calculated by calculating each dot between its weight and a tiny region with which it is associated.

A local extractor feature is used for each kernel matrix set to extract regional characteristics using the input matrices. The learning technique searches for sets of k-kernel matrices, which extract good discriminatory image classification features. In this case, the kernel matrices and biases may be trained with the backpropagation

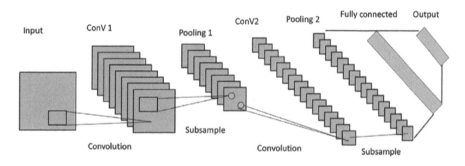

FIGURE 2.2　Convolutional neural network.

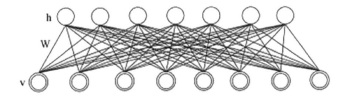

FIGURE 2.3   Illustration of RBM.

approach as shared connection weights, which optimises neural network connection weights.

2.3.3.3   DBN

DBN has emerged as one of the most essential deep learning models. In the pre-training step, it employs a generative model, and in the fine-tuning step, it employs backpropagation. This is beneficial when there is a limited number of training spec-imens, such as hyperspectral remote sensing. DBN is also an algorithm for rapid learning, allowing it to quickly identify the best parameters. We study the efficacy of DBN for hyperspectral data categorisation in this research.

Figure 2.3 shows the top layer with concealed units and the bottom layer with visible units.

RBM is generally utilised in the building of a DBN as a layer-wise training model. It is a network of two layers with "visible" units $v = 0$, 1D and "hidden" units $h = 0$, 1F that shows a certain type of Markov random field (see Figure 2.3). The energy of a combined configuration of the units is given by

$$Ei(v0,h0;\ \theta) = -\sum_{i=1}^{D} bivi - \sum_{j=1}^{F} ajhj - \sum_{i=1}^{D}\sum_{j=1}^{F} wivihj \tag{2.15}$$

$$Pi(v0,h0;\ \theta) = \frac{1}{Z(\theta)}\exp\big(-E(v0,h0;\ \theta)\big) \tag{2.16}$$

$$Zi(\theta) = \sum_{v}\sum_{h} E(v0,h0;\ \theta) \tag{2.17}$$

where $Zi(\theta)$ is the normalising constant. Each input vector is assigned the energy function by the network. Changing the energy given in the training vector may improve the likelihood (1).

The logistic function provides the conditional distributions of hidden unit h and vector $v0$.

$$pi(h0j = 1 \mid v0) = g\sum_{i=1}^{D} Wijvi + aj \tag{2.18}$$

$$pi(vi = 1 \mid h0) = g\sum_{j=1}^{F} Wijhj + bi \tag{2.19}$$

$$g(x) = \frac{1}{1 + \exp(-x)} \tag{2.20}$$

After determining the hidden unit states, each vi with a likelihood of 1 may be reconstructed by the input data (5). The concealed unit status is then modified to reflect the functionality of the reconstruction.

W is learned using a technique known as contrastive divergence (CD). The weight change is provided by

$$\Delta w_{ij} = \epsilon \left(vihj_{\text{data}} - vihj_{\text{reconstruction}} \right) \tag{2.21}$$

where ϵ denotes the learning rate. We may achieve the correct value of W through the learning process. In reconstruction-oriented learning, the power of RBM may be demonstrated. It employs only information that was learned during reconstruction in hidden units such as input features. When the model is able to properly retrieve the initial input, it implies that the hidden units maintain sufficient information about the input and an effective assessment of the input data is possible.

For collecting data properties, the greatest choice is a single covered layer RBM. The characteristics you have learned can be utilised after RBM training as input data for a second RBM. This kind of technology might be used to construct DBN layer by layer. DBN may therefore progressively remove deep features from incoming data. In other words, DBN learns a profound function of the input through training in a hierarchical way. A DBN example related to a future classification is shown in Figure 2.4.

The first RBM converts a first-layer characteristic from the zeroth layer. The course is the same as the previously stated RBM. The follower layers of RBM are learned utilising the output of the preceding layer; the first layer of RBM is completed after training. The learned characteristics of the entire training system are the last characteristics of RBM. An LR layer is inserted at the end of the functional learning system.

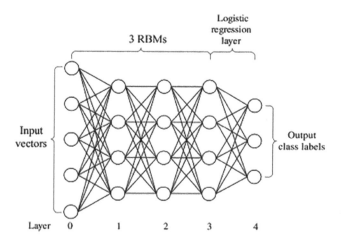

FIGURE 2.4 A DBN instance linked to an LR layer.

This LR classification may be used to fine-tune the whole pre-trained network so that neural network layers can be integrated and classified with learned features. In the peripheral area of parameters, the finishing process, which begins with DBN and seeks a minimum, uses backpropagation.

Algorithm 2: DBN Algorithm

1. Input: the original photograph training set D, prominent subregion size, minimum range of circle and architecture of deep faith network P.
2. The weight matrix W and the pre-trained profound belief network bios are produced.
3. Ddbn = ϕ
4. for each randomly chosen $I0$ $D0$
5. BW IMAGEINBINARIZATION ($I0$)
6. C0← CIRCLEINDECOMPOSITION (B0W0, rmin)
7. $C0$ do for each randomly picked ci
8. $I0$ ←CROPINWINDOW(B0W0, ci, size)
9. Ddbn = Ddbn $I0$
10. Ddbn = Ddbn $I0$
11. end for
12. end for
13. [W0, B0] ← CONTRASTIVEINDIVERGENCE (Ddbn, P0)

2.3.3.4 Random Forest

This section defines our proposed method – the improved random forest algorithm. It is used in big data for improving the artificial intelligence by calculating the resulted image accuracy that is obtained from the above methods such as SVM, CNN and DBN. This resulted calculated values are sorted in ascending or descending order of accuracy.

Random forest is a common model in machine learning since it can be used to deal with both regression and classification issues. It also produces satisfactory results without the need to estimate the hyperparameter. In classification process, random forest tries to enhance the classification value with the use of multiple decision trees. Most important issues in decision-making areas are over-studies, often known as memorisation or, more precisely, overfitting. The forest model chooses and trains tens, if not hundreds, of various sets of data produced randomly from data and characteristics sets, in order to address this challenge. This technology produces hundreds of decision-making bodies, each assessed separately.

The main difficulty with the selection procedure for random forest is how to assess the authenticity of each tree. To measure the importance of a random forest, we used out-of-bag (OOB) accuracy. For the construction of series of training data subsets, the bagging method for random forest model was used; they further create trees. The in-bag (IOB) data are referred to in each tree training subset, whereas the OOB data are the data subset produced from the remaining data. As OOB data do not form trees, the OOB accuracy of each tree can be accessed with it, and also tree's relevance can be judged with this OOB correctness.

In case of a tree classifier $hk(IOBk)$, IOBk and n occurrences in the training data set are formed on the kth training data subset; we define the tree's OOB accuracy $hk(IOBk)$ for each diD as

$$OOBAcck = \frac{\sum_{i=1}^{n} I\left(hk(di) = yi; di \int OOBk\right)}{\sum_{i=1}^{n} I\left(di \int OOBk\right)} \qquad (2.21)$$

In this case, $I(f)$ is an indicator function. The greater the OOBAcck, the better the tree, according to formula (2.21).

The trees are then sorted in order of their OOB accuracies in descending order, with the top 80% trees picked to construct the random forest. A population of "good" trees might arise from this sort of tree selection method. We demonstrate our improved random forest method in this section, which incorporates feature weighting and tree selection approaches. Algorithm 3 lays the groundwork for our methods.

Algorithm 3: High-Performance RF Algorithm

1. Input.
2. Da denotes the training data set.
3. *Ai*: feature space *Ai*1, *Ai*2,…, *AiM*; *Y*: feature space *yi*1, *yi*2,…, *yiq*.
4. *Kt* denotes the number of trees.
5. *mi*: the dimension of subspaces.
6. The result is a random forest.
7. Procedure:
8. Step 1: do for $i = 1$ to *Kt*
9. Step 2: From the training data set Da, create a bootstrap sample IOB data subset IOB*i* and OOB data subset OOB*i*.
10. Step 3: createTree(IOB*i*) = *hi* (IOB*i*).
11. Step 4: Equation (2.14) is used to determine the out-of-bag accuracy OOB Acc*i* of the tree classifier *hi* (IOB*i*) using the out-of-bag data subset OOB*i*.
12. Step 5: end for
13. Step 6: Sort all *Kt* trees classifiers in descending order by OOBAcc.
14. Step 7: Choose the top 80% of trees with strong OOBAcc values and merge the top 80% of tree classifiers to form an enhanced random forest.

This approach includes training data, functional space, class, the number of trees in the RF and subspace size. As a consequence, a random forest model is created. The loop for building K decision trees is formed by Steps 1–5. Step 2 of the loop uses the bootstrap approach to sample the training data, to create the IOB data subset for the tree classifier construction and an OOB data subset for the assessment of the OOB accuracy of the tree classifier. Step 3 invokes createTree to build a tree classification recursive function (). Step 4 employs a data subset to compute the OOB accuracy of the tree classifier. Stage 6 finishes the loop with a decreasing order of the OOB accuracies of all freshly produced tree classifiers. Step 7 chooses the top 80% trees with the largest OOB precision ratings and combines the random forest model with the highest 80% tree classification. In fact, 80% is enough to get good outcomes.

Algorithm 4: Tree Creation Function

1. making new node ni;
2. if the halting requirements are fulfilled,
3. return ni as a leaf node;
4. otherwise
5. from $j = 1$ to $j = M$
6. using equation (2.14), construct the informativeness measure corr(Aj, Y).
7. end for
8. use equation (2.14) to compute feature weights $w1$, $w2$,..., wM;
9. apply the feature weighting approach to choose m features at random;
10. to create an optimum split for the partitioned node, utilise these m characteristics as candidates;
11. for each split, call createTree();
12. end if
13. return ni;

CreateNode() is the method to create a new node. The stop criterion is then used to determine whether the node should be divided or returned to the upper node. The feature weighting technique is used to pick m features at random as a subspace for node splitting while splitting this node. These properties are used to determine the best split for segmenting the node. For each subset of the partition, createTree is called again to create a new node under the existing node. The parent node is returned when a leaf node is generated. This recursive method is carried out until the tree is complete.

This strategy differs from the way of Breiman to generate a random forest model. The first distinction is how every node chooses the subspace of the feature. The fundamental random technique is used by Breiman. In order to incorporate information features in very high-dimensional picture data, the subspace needs to be expanded. As a result, the computing strain is raised. We may still utilise Breiman formula 2 Log () 1 M + for defining the subspace size using the feature weighting technology. The second change is the addition of a mechanism for tree selection. The random forest model enhances that strategy.

2.3.4 EVALUATION

The true-positive rate, false-positive rate and accuracy are used for evaluating the performance of the intrusion detection system. The following equation shows the TPR (2.23):

$$TPR = \frac{TP}{TP + FN} \qquad (2.23)$$

where true positive (TP) denotes the number of invasive samples successfully identified and false negative (FN) denotes the number of invasive samples wrongly identified as benign samples. The detection rate is another name for TPR. The formula for FPR is given in equation (2.24).

$$FPR = \frac{FP}{FP + TN} \qquad (2.24)$$

In cases where FP refers to the number of benign samples that have incorrectly been categorised as invasive, true negative (TN) refers to the accurate number of benign samples.

The false alarm rate is another name for FPR. Equation (2.25) defines the accuracy:

$$Accuracy = \frac{TN + TP}{TP + FN + FP + TN} \qquad (2.25)$$

Table 2.1 and Figure 2.5 shows the accuracy of the Deep Learning classification models such as CNN and DBN.

TABLE 2.1
Accuracy of Deep Learning Classification Models

DATA	CNN	DBN
Data 1	83.2%	79.4%
Data 2	80.1%	82.3%
Data 3	78.8%	74.5%
Data 4	82.3%	77.6%

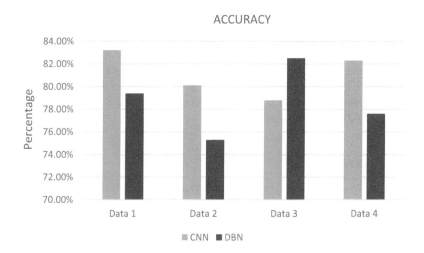

FIGURE 2.5 Accuracy of deep learning classification models.

TABLE 2.2

Accuracy of Machine Learning SVM Classification Model

DATA	SVM
Data 1	88.5%
Data 2	80.6%
Data 3	77.8%
Data 4	87.2%

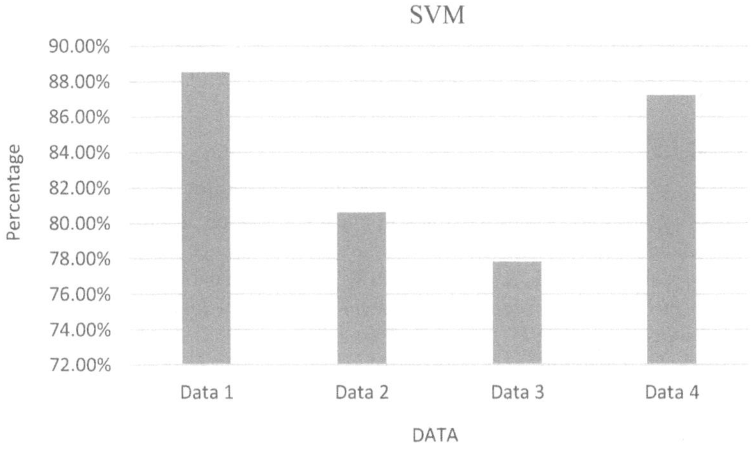

FIGURE 2.6   Accuracy of machine learning SVM classification model.

Table 2.2 and Figure 2.6 show the accuracy of the machine learning classification model SVM.

Table 2.3 illustrates the improved random forest algorithm we proposed in this chapter. This improved random forest classifier is used as an artificial intelligence software in big data for sorting and determining the highest accuracy methods for better and faster image classification and recognition.

2.4   CONCLUSIONS

Deep learning is a rapidly expanding machine learning application. The increasing application of deep learning and machine learning algorithms in several fields demonstrates their success and adaptability. Deep learning's accomplishments and increasing accuracy rates clearly illustrate the technology's use; both deep learning and machine education are emerging well. In this study, we showed the relevance of deep learning in several fields, with picture classification and recognition being one of them. The SVM, DBN and CNN are illustrated for classification of images for better accuracy. The improved random forest algorithm we presented in this chapter

TABLE 2.3

Improved Random Forest Algorithm Classification

DATA	High Accuracy Classifier
Data 1	SVM 88.5%
Data 2	DBN 82.3%
Data 3	CNN 78.8%
Data 4	SVM 87.2%

is depicted in Table 2.3. This improved random forest classifier is utilised in big data as artificial intelligence software for sorting and determining the highest accuracy approaches for better and quicker picture categorisation and recognition.

REFERENCES

1. Dargan, S., Kumar, M., Ayyagari, M. R., & Kumar, G. (2019). A survey of deep learning and its applications: A new paradigm to machine learning. *Archives of Computational Methods in Engineering*, 1–22.
2. Wlodarczak, P., Soar, J., & Ally, M. (2015). Multimedia data mining using deep learning. In *2015 Fifth International Conference on Digital Information Processing and Communications (ICDIPC)* (pp. 190–196). IEEE.
3. Nguyen, G., Dlugolinsky, S., Bobák, M., Tran, V., García, Á. L., Heredia, I., et al. (2019). Machine learning and deep learning frameworks and libraries for large-scale data mining: a survey. *Artificial Intelligence Review*, 52(1), 77–124.
4. Fok, W. W., He, Y. S., Yeung, H. A., Law, K. Y., Cheung, K. H., Ai, Y. Y., & Ho, P. (2018). Prediction model for students' future development by deep learning and tensorflow artificial intelligence engine. In *2018 4th International Conference on Information Management (ICIM)* (pp. 103–106). IEEE.
5. Lin, Y. L., Chen, T. Y., & Yu, L. C. (2017). Using machine learning to assist crime prevention. In *2017 6th IIAI International Congress on Advanced Applied Informatics (IIAI-AAI)* (pp. 1029–1030). IEEE.
6. Wawrzyniak, Z. M., Jankowski, S., Szczechla, E., Szymański, Z., Pytlak, R., Michalak, P., & Borowik, G. (2018). Data-driven models in machine learning for crime prediction. In *2018 26th International Conference on Systems Engineering (ICSEng)* (pp. 1–8). IEEE.
7. Navalgund, U. V., & Priyadharshini, K. (2018). Crime intention detection system using deep learning. In *2018 International Conference on Circuits and Systems in Digital Enterprise Technology (ICCSDET)* (pp. 1–6). IEEE.
8. Zhong, W., Yu, N., & Ai, C. (2020). Applying big data based deep learning system to intrusion detection. *Big Data Mining and Analytics*, 3(3), 181–195.
9. Aksu, D., & Aydin, M. A. (2018). Detecting port scan attempts with comparative analysis of deep learning and support vector machine algorithms. In *2018 International Congress on Big Data, Deep Learning and Fighting Cyber Terrorism (IBIGDELFT)* (pp. 77–80). IEEE.
10. Mohamad, M., Selamat, A., & Salleh, K. A. (2019). An analysis on deep learning approach performance in classifying big data set. In *2019 1st International Conference on Artificial Intelligence and Data Sciences (AiDAS)* (pp. 35–39). IEEE.
11. Aristodemou, L., & Tietze, F. (2018). The state-of-the-art on Intellectual Property Analytics (IPA): A literature review on artificial intelligence, machine learning and

deep learning methods for analysing intellectual property (IP) data. *World Patent Information*, 55, 37–51.

12. Gupta, S., Mohanta, S., Chakraborty, M., & Ghosh, S. (2017). Quantum machine learning-using quantum computation in artificial intelligence and deep neural networks: Quantum computation and machine learning in artificial intelligence. In *2017 8th Annual Industrial Automation and Electromechanical Engineering Conference (IEMECON)* (pp. 268–274). IEEE.

13. Rani, K. S., Kumari, M., Singh, V. B., & Sharma, M. (2019). Deep learning with big data: An emerging trend. In *2019 19th International Conference on Computational Science and Its Applications (ICCSA)* (pp. 93–101). IEEE.

14. Sari, C. T., & Gunduz-Demir, C. (2018). Unsupervised feature extraction via deep learning for histopathological classification of colon tissue images. *IEEE Transactions on Medical Imaging*, 38(5), 1139–1149.

15. wei Tan, J., Chang, S. W., Abdul-Kareem, S., Yap, H. J., & Yong, K. T. (2018). Deep learning for plant species classification using leaf vein morphometric. *IEEE/ACM Transactions on Computational Biology and Bioinformatics*, 17(1), 82–90.

16. Chandran, P. S., Byju, N. B., Deepak, R. U., Nishakumari, K. N., Devanand, P., & Sasi, P. M. (2018). Missing child identification system using deep learning and multiclass SVM. In *2018 IEEE Recent Advances in Intelligent Computational Systems (RAICS)* (pp. 113–116). IEEE.

17. Dong, Y. N., & Liang, G. S. (2019). Research and discussion on image recognition and classification algorithm based on deep learning. In *2019 International Conference on Machine Learning, Big Data and Business Intelligence (MLBDBI)* (pp. 274–278). IEEE.

18. Sufri, N. A. J., Rahmad, N. A., Ghazali, N. F., Shahar, N., & As'Ari, M. A. (2019). Vision based system for banknote recognition using different machine learning and deep learning approach. In *2019 IEEE 10th Control and System Graduate Research Colloquium (ICSGRC)* (pp. 5–8). IEEE.

19. Dong, Y., Zhang, Q., Qiao, Z., & Yang, J. J. (2017). Classification of cataract fundus image based on deep learning. In *2017 IEEE International Conference on Imaging Systems and Techniques (IST)* (pp. 1–5). IEEE.

20. Huixian, J. (2020). The analysis of plants image recognition based on deep learning and artificial neural network. *IEEE Access*, 8, 68828–68841.

21. Mete, B. R., & Ensari, T. (2019, October). Flower classification with deep CNN and machine learning algorithms. In *2019 3rd International Symposium on Multidisciplinary Studies and Innovative Technologies (ISMSIT)* (pp. 1–5). IEEE.

22. Zawbaa, H. M., Hazman, M., Abbass, M., & Hassanien, A. E. (2014). Automatic fruit classification using random forest algorithm. In *2014 14th International Conference on Hybrid Intelligent Systems* (pp. 164–168). IEEE.

23. Sewak, M., Sahay, S. K., & Rathore, H. (2018). Comparison of deep learning and the classical machine learning algorithm for the malware detection. In *2018 19th IEEE/ ACIS International Conference on Software Engineering, Artificial Intelligence, Networking and Parallel/Distributed Computing (SNPD)* (pp. 293–296). IEEE.

24. Badriyah, T., Sakinah, N., Syarif, I., & Syarif, D. R. (2020). Machine learning algorithm for stroke disease classification. In *2020 International Conference on Electrical, Communication, and Computer Engineering (ICECCE)* (pp. 1–5). IEEE.

25. Lee, K. B., & Shin, H. S. (2019). An application of a deep learning algorithm for automatic detection of unexpected accidents under bad CCTV monitoring conditions in tunnels. In *2019 International Conference on Deep Learning and Machine Learning in Emerging Applications (Deep-ML)* (pp. 7–11). IEEE.

26. Kavitha, D., Hebbar, R., Vinod, P. V., Harsheetha, M. P., Jyothi, L., & Madhu, S. H. (2018). CNN based technique for systematic classification of field photographs. In

2018 International Conference on Design Innovations for 3Cs Compute Communicate Control (ICDI3C) (pp. 59–63). IEEE.

27. Li, Q., Cai, W., Wang, X., Zhou, Y., Feng, D. D., & Chen, M. (2014). Medical image classification with convolutional neural network. In *2014 13th International Conference on Control Automation Robotics & Vision (ICARCV)* (pp. 844–848). IEEE.

28. Akila, D., Pal, S., Jayakarthik, R., Chattopadhyay, S., & Obaid, A. J. (2021). Deep learning enhancing performance using support vector machine HTM cortical learning algorithm. *Journal of Physics: Conference Series*, 1963, 1, 012144.

29. Suseendran, G., et al. (2021). Deep learning frequent pattern mining on static semi structured data streams for improving fast speed and complex data streams. In *2021 7th International Conference on Optimization and Applications (ICOA)*. IEEE.

30. S. Lavanya; D. Akila, Prediction performance of crime against women using rule based decision tree J48 classification algorithms in various states of India. *IEEE Xplore*, 139–144.

3 Implication of Soft Computing and Machine Learning Method for Software Quality, Defect and Model Prediction

Anurag Sinha
IGNOU

Shubham Singh
Galgotias University

Devansh Kashyap
Kalinga Institute of Industrial Technology

CONTENTS

DOI: 10.1201/9780367816414-3

3.1 INTRODUCTION: OVERVIEW OF THE STUDY

In terms of software engineering, software quality refers to the performance quality of the program and the quality of the program structure. The quality of the program reflects the working conditions, while the quality of the structure emphasises the non-functional requirements. Software estimates focus on production, process and project quality aspects. In this chapter, the most important focus is on the software invention. The reason of software excellence work is to realise the requisite structure excellence by defining and implementing quality requirements, measuring proper eminence attribute and evaluating the resulting quality of the software quality. The measured changes done in software defects have certain standard feature and components (appropriate). This can be done in the form of quality or quantity, or a combination of both. In both cases, each desired feature has measurable features, such as application design standards, coding methods, complexity, documentation, portability and technical and performance capabilities. The existence of these features as part of a software or system seems to be related to or associated with this feature (Sinha et al., 2020).

Software defects recognition plays a significant function in vigorous research in software engineering. Software bugs are software defects; errors in program code; and mistakes or blunders that result in off-base or surprising outcomes. The identification of significant danger factors related to programming disappointments, which were not recognised in the beginning phases of programming improvement, is tedious. Mistakes can happen at any phase of programming improvement. Rising programming organisations are zeroing in on programming quality, particularly in the beginning phases of programming advancement. In this manner, the fundamental objective of every association is to distinguish and dispense with defects in the early life cycle development program (SDLC). Information mining methods are utilised to work on the nature of the program and to make forecasts about programming devices utilising verifiable information and bugs. This chapter furnishes the reader with transient information mining methods, taking gander at the most recent improvements in determining in the defected portion of code block.

Programming techniques give various devices to programming improvement and quality control of program creation. It is important to decide the assets required, which are significant source of dynamic data. Various advances have been proposed in the writing. Its done on the estimation construction to help artistic expressions against quality previously during the improvement interaction. Plan quality evaluation is level headed, and estimations can be programmed. Yet, how would we know which exercises are truly in the vital quality perspectives? The ISO/IEC standard (14598) states that inward markers are especially valuable with regard to outside quality attributes, such as consistency and reuse. Various methodologies have been proposed to foster more refined assessment models; for instance, they can be numerical models (on account of measurable strategies direct and calculated returns) or man-made consciousness models (on account of AI procedures). So our work is identified with the improvement of compelling and/or reasonable component assessment models, particularly manageability and reuse. We utilised distinctive Machine Learning calculations to make these models. In this review, we are keen on computing their presentation and assessing their comprehension of computer programming information. Execution alludes to quantitative estimation, which is regularly communicated in the exactness of the model, while appreciation calls for explanation and comprehension of the model. In this way, in Section 3.2, we first present the distinctive Machine Learning calculations we use. In Section 3.3, we depict the authentic cycle we follow, and then, at that point, introduce and examine the created models as far as execution and comprehension. At long last, Section 3.4 sets out the ends. Lately, software engineers (SEs) have zeroed in on information mining (DM) and AI (Machine Learning) in light of examination, as SE information assortment can assist with discovering new data. Computer programming offers a wide scope of exploration subjects, and information investigation can give extra bits of knowledge to help dynamic subject area. Figure 3.1 shows the crossing points of the three fundamental parts: information mining, programming improvement and measurements. A ton of information is gathered from associations during programming improvement and

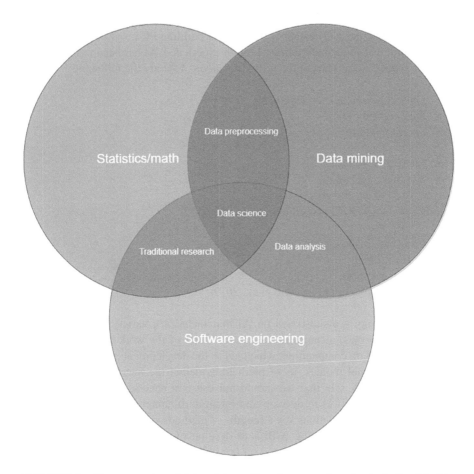

FIGURE 3.1 The connection of data mining with software engineering including Machine Learning.

support exercises, such as portrayal definitions, plan charts, source numbers, blunder reports and program types. Information assortment gives valuable information and concealed techniques from SE information. Maths gives essential capacities, and insights decide the likelihood, relationship and connection of accumulated information. Information science at the core of the graph covers an assortment of disciplines, such as DM, SE and insights. The review gives an outline of how to manage SE issues utilising diverse extraction techniques. Thus, multi-criterion based software system utilised the ANN configuration arranging arrangement depict the instruments and procedures regularly utilised by specialists checked learning issues. In initial stage team of software engineers validated the Machine Learning model for software defect prediction.

3.2   BACKGROUND: MACHINE LEARNING
FOR DEVELOPING MODELS

Artificial intelligence (AI) manages the issue of building PC programs that work on their exhibition at some undertaking through experience. AI has been used in different issue spaces. Some runs of the mill utilisations of AI are the following:

- Optical person acknowledgment
- Face recognition
- Spam separating
- Fraud recognition
- Medical conclusion
- Weather expectation.

Significant classes of AI procedures are as follows:

- Case-based calculation
- Rule initiation
- Neural networks
- Genetic algorithm (Tejaswini et al., 2019).

Machine Learning is a site based on AI that includes a variety of achievements (meetings, magazines, techniques and tools). Much of the work done on Machine Learning focused on computation, which led to the development of various systems. By drawing fictional models, these predictions define the meaning of each class. The choice of study size that we should not use is a necessary development. It's certainly worth the effort to develop and integrate innumerable systems. Decision trees and rules and regulations key strengths of methods such as Bayesian networks (BNs) are case-based learning (CBL), to develop models that we can involve in decision-making. It is a system based on fair classification of data between explicit data (examples, rules, trees and strategies that validate these points). Methods such as astronomical networks (ANNs and manipulators (SVM)) are considered "concepts". As part of the initial disclosure statement (organisation, design, fractions), we have no other job to handle them and expect production once they have been prepared. Understanding balance is one of the most likely uses of ANN design to solve project problems in control learning. Here are some suggestions on how to look or get an appointment for acne treatment in one of the following ways: Acute back pain (RBP) is one (Challagulla et al., 2005). Finally, SVM often conducts introductory training to address two design validation issues. SVM provides training and inspection; efforts have been made to accumulate efforts such as quality control and other things. The only philosophy/strategy is to find a compromise between the features, and we need to complete all the information. The participating states will present purchase forms using data related to certain product quality criteria. They will discuss their abilities and prosperity (Wahono, 2007).

Evaluating the forecast execution of a classifier is in view of the disarray lattice (see Table 3.1, where the cells include frequencies for every blend of the two paired dichotomous factors). Overall, the implications of the upsides of the double factors need not be characterised; be that as it may, for evaluating grouping performance, we are more explicit. The class names are named positive and negative. It is standard to utilise the positive name to allude to the result of interest, so in our circumstance, the positive mark eludes to the product part being inadequate. Subsequently, we have several performance metrics to evaluate our model are those examples that the classifier mistakenly doles out to the imperfection inclined class, etc. Deciding characterisation execution is more unpretentious than it may initially show up since we need to take into account both possibility parts of a classifier (in any event, speculating can prompt some right arrangements) and furthermore what are named lopsided informational collections where the dissemination of classes is a long way from 50:50. As examined beforehand, this is an ordinary circumstance for deformity informational collections since most programming units don't contain known imperfections. Many regularly utilised measures, for example the F-measure, are unacceptable because they are not based on the total disarray lattice. A generally utilised alternative is the area under the curve (AUC) of the ROC graph; not withstanding this, since this is an action on a family of classifiers, it can't be deciphered except if one classifier stringently overwhelms since we're not given the general expenses of FP and FN. As such, for two classifiers A and B,

$$AUCA > AUCB = 6 \Rightarrow A - B \tag{3.1}$$

TABLE 3.1
Precision Measurements

Term	Formula	Definition
True-positive pace	$Tp/(tp+fn)$	Amount of defective units perfectly classified
Precision	$Tp/(tp+fp)$	Part of unit precisely predicted as imperfect
F-measure	$2*$ Recall precision$/tp+fp$	Common factor for recall and precision
Accuracy	$tn+tp/tn+fn+tp+fp$	Amount of acutely classified part

For this explanation, we advocate a paired connection coefficient differently known as the Matthews connection coefficient (MCC). Unlike the F-measure, MCC is based on each of the four quadrants of the disarray grid. MCC is indistinguishable from the parallel relationship coefficient φ initially because of Yule and Pearson in the mid-20th century. In any case, we will allude to the action as MCC instead of φ since we are managing the specific parallel factors and implications of anticipated the more, genuine class though the φ coefficient is a more general proportion of affiliation, where the translation of a negative connection is harder to determine. It is a fair measure and handles circumstances where the proportion of class sizes are profoundly imbalanced, which is ordinary of programming imperfection information (classes containing absconds are frequently generally uncommon). A zero worth demonstrates the two factors are autonomous: Tending towards solidarity demonstrates a positive connection among anticipated and real classes and tending towards less solidarity a negative relationship. Strangely one could change over an extremely poor or, on the other hand, unreasonable classifier into an awesome one by nullifying the forecast (turning a positive to negative or bad habit, vice versa). At long last, MCC can be effectively registered from the disarray lattice.

3.3   RELATED STUDY

There are incredible assortments of studies which have applied factual and AI-based models for imperfection forecast in programming frameworks have utilised calculated relapse to analyse what the impact of the set-up of article situated plan measurements is on the forecast of issue inclined classes have utilised the neural organisation to characterise the modules of huge media transmission frameworks as shortcoming inclined or not furthermore, contrasted it and a non-parametric discriminate model (Tantithamthavorn et al., 2016). The results of their examination have exposed that in contrast to the non-parametric discriminate model, the prescient exactness of the neural organisation model had a superior outcome. Then, at that point, Laradji et al. (2015) presented a defence concentrate by utilising relapse trees to arrange deficiency inclined modules of huge telecom frameworks. They utilised Bayesian Belief Organization to distinguish programming surrenders. Nonetheless, this AI calculation has heaps of limits, which have applied random timberland calculation on programming deformity dataset presented by NASA to anticipate deficiency inclined modules of programming frameworks. Pelayo and Dick (2007) also contrasted their model and some measurable and Machine Learning models. The results of this correlation showed that in contrast to different strategies, the calculation gave better precision for the proposed a model, which utilises three AI calculations: decision tree, multilayer perceptron and outspread basis functions, to recognise the effect of this model to foresee on various programming metric datasets obtained from the real-life tasks of three major size programming organisations in Turkey. The outcomes showed that limitations apply to the entirety of the AI calculations had comparable outcomes which have empowered to anticipate possibly damaged programming furthermore, make moves to address them have researched the effect of support vector machines (SVMs) on four NASA datasets to foresee imperfection inclination of programming frameworks and looked at the expectation execution of SVM against eight measurable and AI models. The results demonstrated that the expectation execution of SVM was far superior to other people have explored the effect of the commotion on imperfection expectation to adapt to the commotion in imperfection information by utilising a clamour identification and disposal calculation (He et al., 2015). The results of the investigation introduced that boisterous cases could be anticipated with sensible exactness, and applying disposal improved the deformity expectation exactness. They researched re-inspecting methods, outfit calculations and limit moving as class lopsidedness learning techniques for programming imperfection forecast. They have utilised various strategies; among them, AdaBoost.NC would do well to evaluate expectation execution. They additionally worked on the viability and effectiveness of AdaBoost.NC by utilising a unique adaptation of it and proposed a model to settle the class irregularity issue, which causes a decrease in the execution of deformity forecast. The Gaussian capacity has been utilised as bit work for both the asymmetric kernel incomplete least squares classifier (AKPLSC) and asymmetric portion principal component analysis classifier (AKPCAC); what's more, NASA and SOFTLAB datasets were utilised in testing. The outcomes showed that the AKPLSC had better effects on recovering the misfortune brought about by class awkwardness; what's more, the AKPCAC would be used to foresee abandonment on imbalanced datasets (Nam, 2014).

3.4 LITERATURE REVIEW

There are various evaluations about programming bug doubt using AI strategies. For example, the evaluation using the proposed straight regression approach expected broken modules. The evaluation predicts future deficiencies depending on the chronicled data of the thing amassed issues. The assessment, moreover, looked into and isolated the AR model and the known power model of Machine Learning used for root-mean-square error assessment. In any case, the appraisal used three datasets for assessment and the results were promising. The assessments reviewed the relevance of various Machine Learning systems for deficiency doubt. Rawat and Dubey (2012) added to their appraisal the most fundamental past explorations about each Machine Learning procedure and the most recent things in programming bug assumption using AI. This appraisal can be used as the ground or a step to anticipate future work in programming bug assumption. Rodriguez et al. (2014) presented a decent cognisant review for programming bug assumption, taking a look at strategies using Machine Learning. The paper interweaved a layout of the gigantic number of studies between the years 1991 and 2013, destroyed the Machine Learning techniques for programming bug speculation models, reviewed their show, checked out among Machine Learning and appraisal methodologies, pondered between different Machine Learning procedures and summarised the strength and insufficiency of the Machine Learning systems. Li et al.'s (2012) paper gave a benchmark to allow a typical and obliging assessment between different bug assumptions moves close. The evaluation presented a total association between a striking bug assumption moves close and additionally introduced a new strategy and surveyed it by building a good examination with various procedures using the presented benchmark. Dam et al. (2018) enabled a model for object-facilitated software bug prediction system

(SBPS). The evaluation merged identical sorts of blemish datasets that are open at software engineering repository. The evaluation concentrated on the proposed model by using the measure (precision). Finally, the examination results showed that the ordinary proposed model exactness is 78.2%. The application gets its characteristics, for instance, the thing arranged evaluations and count appraisals regards from an open-source programming project. The innate evaluation uses the application's ascribes as liabilities to pass on rules which used to sort the thing modules to harmed and non-imperfect modules. Finally, imagine the yields using genetic evaluation applet. The evaluation in Sun et al. (2012) reviewed by used AI procedures (decision tree and neural affiliations) and certifiable methodology (reliable and organise lose the confidence). The coupling between object (CBO) metric is the best to evaluate the bugs in the class and the line of code is all over well, yet the depth of inheritance tree and number of children are unreliable estimates. Zheng (2010) researched five standard Machine Learning estimations used for programming mutilation doubt, which are fake neural networks, particle swarm optimisation, decision tree, naïve Bayes and linear classifiers. The assessment presented fundamental results, including that the ANN has the least screw up rate followed by DT, but the straight classifier is better than various computations in terms of blemish actually look at accuracy. The most standard methods used in programming event speculation are DT, BL, ANN, SVM, RBL and EA, and the normal evaluations used in programming imperfection assumption contemplates are line of code (LOC) appraisals and object coordinated appraisals such as connection, coupling and heritage. Similarly, various appraisals called cross-breed appraisals used both article facilitated and procedural appraisals. In the same manner, the results showed that most programming deformation doubt considered used NASA dataset and promise dataset. Moreover, the evaluations in Arora et al. (2015) took apart indisputable Machine Learning approach and gave as far as possible in programming distortion doubt. The evaluations helped the fashioner with using tremendous programming appraisals and fitting data mining system to update the thing quality. The examination picked the best appraisals that are useful in disfigurement doubt, such as response for class, line of code and lack of coding quality. Faint et al. (2011) presented the most exceptional data mining system. The appraisal inspected and thought about four estimations and took apart the advantages and impedances of each evaluation. The conceded results of the examination showed that there were different parts impacting the exactness of each technique, such as the shot at the issue, the used dataset and its affiliation. Amershi et al. (2019) presented the connection between object-arranged appraisals and deficiency tendency of a class. They showed that the algorithms are useful in expecting gives up; in the same way, they showed that the AUC is a reasonable estimate and can be used to predict the damaged modules in the early phases of programming progress and to deal with the precision of Machine Learning methodology. The paper neatly surveys the Machine Learning classifiers using specific execution examinations (for instance, exactness, accuracy, audit, *F*-measure and ROC wind). Three public datasets are used to outline the three Machine Learning classifiers. Of course, a huge amount of the implied related works analysed more Machine Learning frameworks and evident datasets. A piece of the past appraisals incredibly loped around the appraisals that make the system as capable as could really be anticipated, while other past assessments proposed different methodologies to expect programming bugs as opposed to Machine Learning frameworks (Rahman et al., 2019).

3.5 METHODICAL REVIEW: SOFTWARE DEFECT PREDICTION USING MACHINE LEARNING

A software bug is a bug or imperfection. There may not be any program or PC system that produces wrong or amazing results, nor may it appear suddenly. Most bugs affect the structure of the program; its structure; or because of errors and factors in the structures and operating systems used in this job, and the coordinators who created the wrong code. The plane error prediction model can attract these models and try to anticipate programs that include experiences like a desert. There is a correlation between object rating and object orientation. Types of software incapacity metrics include independent components (software metrics) that include lifecycle development programs and component assessments (inadequate or non-invasive). There are different ways to find great systems. Data mining is a testament to man-made mental capacity. This is the assessment process of the "databases" cycle, which aims to collaborate on a variety of educational programs, including certain knowledge and data collection. The overall goal of data mining is to extract data from a study file and turn it into a logical diagram for further investigation. Mining information can be divided into two types: forecasting activities and modelling efforts. The current task is to estimate the exact value of the quality (target/variable) according to the value of the different titles (legitimacy). The drawing effort involves determining blueprints (organisations, examples, and perspectives) that summarise the secret relationship between the data. There are in-depth knowledge-based techniques for the following explored programming hypotheses (Zanutto et al., 2012).

1. **Regression Model**: It is a measurable cycle to assess the connection between factors. It monitors the relationship between the variable or component variable and self-owned or indicator factors. The relationship is conveyed as a condition that predicts the response variable as an immediate limit of pointer variable.
2. **Association Rule Mining**: It is a strategy for finding intriguing connections between factors with regard to huge information bases. It is tied in with discovering affiliations or connections among sets of things or items in a data set. It essentially manages discovering decisions that will foresee the event of thing dependent on the event of different things.
3. **Clustering**: Clustering is an approach to order an assortment of things into gatherings or groups whose individuals are comparable here and there. It is assignment of collection a bunch of things so that things in a similar group are like one another and unlike those in different bunches.
4. **Classification**: It comprises foreseeing a specific result dependent on given information. The order method utilises input information, also called preparing set where all articles have as of now been labelled with realised class marks. The intended inference result is to break down and gains from the

preparation informational collection and fosters a model. This model is then used to order test information for which the class names are not known. The different characterisation strategies are given underneath.

5. **Neural Networks**: These are simple models that can be achieved through the preparation and reception of natural nervous systems in the structure. A nervous system is made up of interconnected nerves that work on the inside as well.

6. **Decision Trees**: A decision tree is a very smart design that can be used to design both a game plan and a step back in tree design. It points to the reformists' decisions and their results. It is a tree with decision centres and leaf centres. A decision-making body has two divisions. Leaf centres process a request or a decision.

7. **Naive Bayes**: It depends on Bayes hypothesis with freedom suspicion between indicators. The innocent Bayes classifier depends on the understanding that the presence or non-appearance of a specific component of a class is not identified with the presence or non-attendance of some other elements.

8. **Support Vector Machines**: A SVM depends on the probability that the decision plane expresses decision constraints. The decision maker is one of the few disconnected items in the classroom. It is essentially a classifier procedure that creates a multidimensional space that differentiates the symbols of each class and performs the query task. It maintains both regression and aggregation. Case-based reasoning: Case-based reasoning uses old problems to deal with new problems and explain new situations. It works by distinguishing between new unclassified records and popular models and models. The direct conclusion of a case-based learning estimate is the computation of the nearest neighbour of k. It directly calculates the storage of each available case and the collection of new subjects for an equivalence measure for distance examples (Ahmed et al., 2020).

3.5.1 Approach of Software Defect Prediction

For the most part, three methodologies are performed to assess forecast models. Cross-marking projection for mixed data collection.

One form of prediction of defects within the project can be created with data events that can be verified by a draft article and is called IPDP, which predicts deficiencies in a comparison effort. The program has access to properly documented information. The points of the model are that faulty measurement areas are usually existing neighbourhood data (for example, in the concept of project distortion) that an organisation needs to focus on data planning; pay attention to project estimates and related information from previous projects. The work best in projects as long as there is some interesting information to design the layout. This means that we have to start with the obvious facts to improve the distorted brand. If you lose data, you can use a different one. The in-in project's distortion gauge loads make it impossible to purge these chronic data 100% accurately using IPDP. Yes, recorded data are usually not shown for retry and some correlation. This hampers the idea of a strong loophole in the current situation. To solve this problem, we used the software Error Prediction Framework (Feldt et al., 2018).

IPDP attempts to establish a uniform standard for all IPDP data to be used in one form or another without objective data. Therefore, an estimate was made for one attempt and then it was applied to another attempt or project. For example, start with an effort and watch measurement patterns in motion, and then move on to the next task. The disadvantages of applying the IPDP include the need for estimates from similar projects; overall estimates are the results that should be compared to projects. Therefore, the current IPDP systems are confusing to link to different datasets and projects (Pradhan et al. 2020).

Differentiated estimates are used to manage the insufficient use of comparable data for IPDP for different data sets; in project defect prediction, it is possible to predict errors in a similar process by collecting data that can be verified from a construction model. Project defect estimation works best if there is enough authentic information available to modify project defect estimation forms. 45. Product anomalies exist in the form of incomplete estimates, and an organisation must have a database to implement information nearby (for example, during project failure estimates); protests related to the measures and weaknesses of the project were eliminated. As long as there is satisfactory information to change the design, skewed predictions indicate that it works best on the project. This means that we have to accept the recorded information in order to improve the distorted index. If data are missing, cross-company defect prediction (CCDP) can cause project imperfections. It is not possible for all companies to continuously collect such verifiable information. For new businesses and some organisations, information that is not disclosed often is not often introduced. An effective failure prediction for this situation is confusing. To overcome this problem, go beyond distorted vision (Washizaki et al., 2019).

Software Defect Prediction Techniques: To chip away at the suitability and nature of programming improvement and to expect gives up in programming, distinctive data mining strategies can be applied to different software engineering areas. The thoroughly used SDP techniques are data mining methodologies and AI methodology and are depicted in Figure 3.2.

From Figure 3.2,

Supervised Learning: One categorical group proposes to address the problem of inconsistent programmatic imperfections. This model depends on the appropriateness integration (APE) approach. Collective learning is a way to collect patterns from grassroots groups. Such models are typically used to demonstrate information drawbacks and antitrust rules that prevent the programmer from collecting imperfect data. This strength is confirmed by performing the characterisation of many classifiers. This average result eliminates unrelated errors and subsequently improves the performance of the group in general. Unlike the voting methods, the probability results are the same as the AUC scale, which assesses the level of certainty associated with the selected class. In addition, a selection of classes requires a restriction on the choice of classes. This constraint can be checked as a

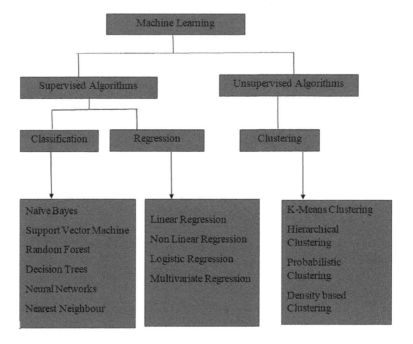

FIGURE 3.2   Machine Learning algorithms.

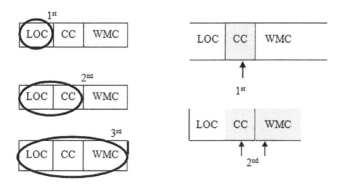

FIGURE 3.3   Forward and backward selection.

quantisation step to signal errors in options. In this way, our determination of a normal probable team meets two requirements. (i) Consistency with the AUC scale and (ii) the narrow selection requirement is measured on definite point (Rana and Staron, n.d.) (Figure 3.3).

Random Forests: These comprise of a few unprimed order or relapse trees. Utilising irregular element determination, these trees are initiated from bootstrap tests of the preparation information [20]. In grouping issues, every information test is taken care of down every one of the trees in the arbitrary backwoods. Then, at that point, the last yields as its choice class the class that got the majority of the votes made by the singular trees. It is shown that blunder rates in irregular backwoods rely upon the strength of every individual tree and the relationship between any two trees in the woodland. Be that as it may, results separated from arbitrary backwoods are hard to decipher. A run of the mill irregular woods is displayed in Figure 3.4. In such settings, every individual tree handles a little subset of elements chosen arbitrarily. Then, at that point, each tree is advanced utilising this subset.

Gradient Boosting: Boosting takes care of relapse issues utilising an expectation model comprising a group of frail indicators [49]. These indicators are normally choice trees. Given a bunch of choice trees $T1$, $T2$, $T3...TN$, the angle boosting calculation creates a weighted summation of the yield choices of every individual tree as follows (Xing et al., n.d.):

$$f(x) = w0 + w1\,h1(x) + w2h2(x) + ...+ wnhn(x) \tag{3.2}$$

Stochastic Gradient Descent: Arrangement and relapse issues including huge datasets are effectively addressed utilising second request stochastic slope and found the middle value of stochastic inclination procedures [50]. In the stochastic angle plunge, cost capacities are limited utilising the stochastic model (SGD):

$$wk + 1 = wk - \{\mu\, Ww\, Q(xk, wk)\} \tag{3.3}$$

Calculated Regression: The calculated relapse gives an exceptionally incredible discriminative model dependent on the notable strategic (sigmoid) work. The calculated capacity, displayed in Figure 3.5, has extremely alluring properties, including ceaseless differentiability and direct connection between the capacity and its subordinates (of any request) [22]. The strategic relapse has effectively been applied in characterisation issues. Given two classes, named $Y=0$ and $Y=1$, and n-dimensional components $\{x1, x2, ..., xi,..., xN\}$ where each element test is treated as an irregular vector comprising discrete arbitrary variable, the strategic relapse yields a generative model that learns $p(Y|x)$ utilising an immediate utilisation of Bayes rule as follows:

A normal flow of defect prediction using Machine Learning algorithm is as follows:

Marking: Defect information ought to be accumulated for preparing a forecast model. In this cycle normally extricating of occasions for example information things from programming chronicles and marking (TRUE or FALSE) is finished.

Removing Highlights and Making Preparing Sets: This development involves the development of provisions for predicting used brands. To predict deformation, it is generally confusing: screams and shouts. It measures changes and background conditions. By combining symbols and objects, we can provide a preparation for a mechanical student to build a prediction model.

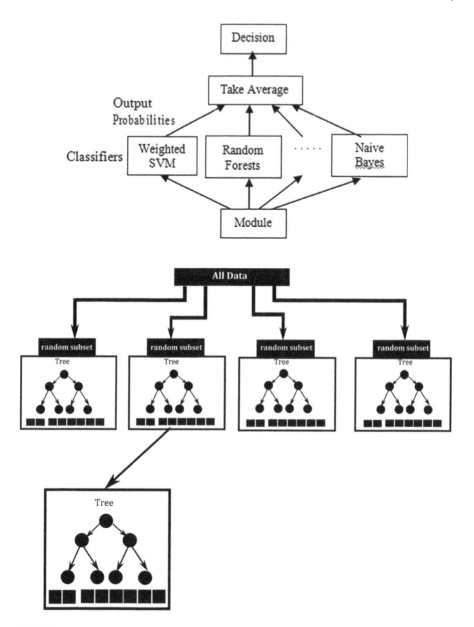

FIGURE 3.4   Random forest algorithm mechanism.

Building Prediction Models: For example, part of general machines can use SVMs or the Bayesian Network to create a vision model using a set of configurations. The model can take another example and predict its brand, for example Valid or false.

Appraisal: The evaluation of an estimate requires an experimental information index, except for a preparation. Signs that occur early in the experience

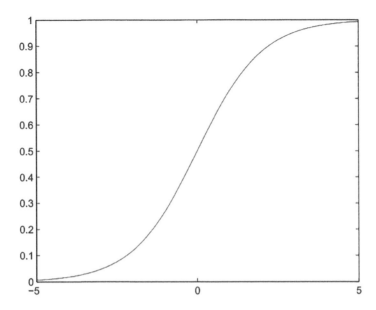

FIGURE 3.5 Sigmoid function graph of regression.

are anticipated and evaluated by comparing expectations with actual names. Full ten-layer approval is fully used to separate prep and test package (Rashid, 2012) (Figure 3.6).

3.5.2 DEFECT PREDICTION BY SOFT COMPUTING METHOD

The Probabilistic Model for Defect Prediction Utilising Bayesian Belief

Network: Probabilistic model for deformity expectation. They suggested a comprehensive model instead of a solitary issue (for example size, or intricacy, or testing measurements, or interaction quality information) model, by consolidating the various elements of easy-going proof to effective imperfection forecast. The model uses Bayesian belief network (BBN) as the reasonable practice for portrayal of this proof. The Bayesian methodology makes measurable end be improved by master judgment in those pieces of an issue circle where exact information is dissipated. Also, the causal or impact association of the model better mirrors the series of true occasions and relations than some other practice. BBN can be taken advantage of to help powerful dynamic for SPI (Software Process Improvement), by executing the accompanying advances (Figure 3.7).

Fuzzy Logic Approach: The fuzzy logic model depends on the idea or thinking and deals with a worth that is inexact in nature. It is a move forward from regular Boolean logic where the value must be TRUE or FALSE. If there should arise an occurrence of fuzzy rationale, the reality of any assertion is degree and not an outright number. Displayed on human instinct and conduct, the greatest in addition to point of fuzzy rationale is that instead of the

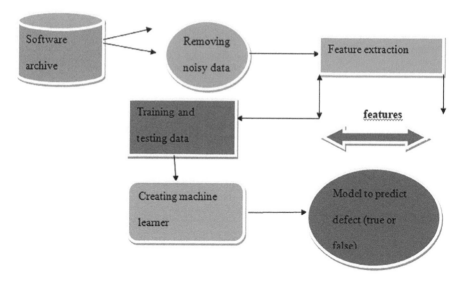

FIGURE 3.6 Framework for defect prediction.

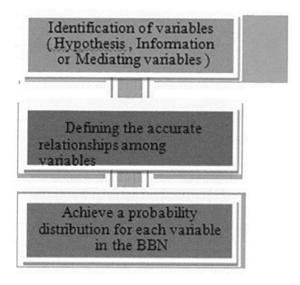

FIGURE 3.7 Bayesian approach for defect prediction.

customary yes–no replies, this model components in the level of truth and subsequently makes portion for the more human-like replies. This model uses data sources and places them in a reach framework. After this, a bunch of decisions are characterised, which direct and impact how sources of info will be used in getting the yield just as tracking down the authoritative worth in the fluffy set. The model has a bunch of measurements or unwavering

quality applicable measurement (Machine Learning) list, which is produced
using the accessible programming measurements. The measurements are
appropriate to their separate stages in the product advancement life cycle
(Tejaswini et al., 2019).

Necessity Phase Metrics: As you can see, the model has utilised three prerequisite measurements (RMs), which are prerequisites Change Request;
Review, Inspection and Walk Through; and Process Maturity (PM) as contribution to the necessities stage.

Configuration Phase Metrics: Like the above stage, three plan measurements, for example configuration imperfection thickness, shortcoming days
number, and information stream intricacy, are taken as info.

Coding Phase Metrics: In this stage, two coding measurements, for example
code deformity thickness and cyclamate intricacy, are taken as contribution at coding stage. The yields of the model will be the quantity of issues
towards the finish of requirements phase; the number of faults towards the
finish of design phase; and the number of faults towards the finish of coding
phase (Figure 3.8).

Defect Prediction Models Based on Genetic Algorithms: Hereditary algorithms are a way to deal with AI, which acts also to the human quality and
the Darwinian hypothesis of regular determination. They are part of the evo-
lutionary algorithms that create arrangements dependent on the strategies
all the more normally found in nature, such as change, determination and
hybrid. Hereditary algorithms are executed starting with a singular populace
that is normally addressed as trees. A potential arrangement is addressed by
each tree, or say chromosome, for this situation. Hubs on the tree imply specific qualities that identify with the issue for which the arrangement is being
looked. All things considered, the arrangement of possible answers for the
issue is (addressed by the chromosomes) known as the populace.

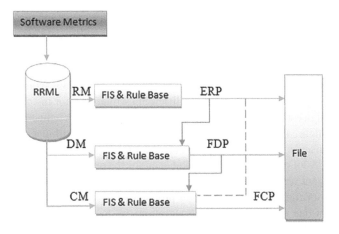

FIGURE 3.8 Fuzzy logic approach.

In the first place, genetic algorithms start with a huge populace. In that populace, every individual addresses a conceivable answer for the issue. These people in the populace are then encoded in a double string that is known as a chromosome. From that point forward, the gathering of the people will contend so they can imitate and afterwards define the future. Nonetheless, there is a capacity called the wellness work that figures out which of the contending people will acquire the option to imitate. Having the wellness work set up ensures that hands down the best people of the populace will actually want to extend their posterity into the future. The cutting edge is shaped by the accompanying exercises occurring.

   a. **Reproduction**: The proliferation measure happens when two chromosomes trade a piece of their code to shape the new people. The hybrid focuses (where the pieces of the code will trade) are chosen by irregular (for a basic rendition of the calculation). At the hybrid point, the chromosomes trade the information keeping the first information up to that point.
   b. **Mutation**: This comes in to present variety in the cutting edge which forestalls the coming to of neighbourhood minima. While the hybrid adjusts the qualities after a haphazardly chosen hybrid point between two chromosomes, transformation chooses a hub in the tree of one chromosome and changes the hereditary material.

This cycle rehashes the same thing until there is an ideal arrangement set came to (ideal wellness level). Be that as it may, there are events when this doesn't occur. In such cases, the program ends after a bunch of emphases. The emphases of the returns are otherwise called ages (Azar and Vybihal, 2011) (Figure 3.9).

3.5.3 DATA MINING IN IMPERFECTION EXPECTATION

An error is an error in a system that causes fundamental or miraculous results. It means an imperfection or a defect. Because the quality of the programmer goes down to the defect of the object, an object should not be distorted. However, projects require a lot of time or people planning to get rid of them before they throw something away. In this case, bug fixes can help recognise and eliminate executions in the early days of SDLC and create surprisingly responsive programming structures. Therefore, the Programming switch predicts programming fault allows all progressive programming structures to be reduced. Various examinations have been driven on blemish estimate using different estimations, for instance, code multifaceted design estimations, object-arranged estimations and connection estimations to assemble figure models. These models can be considered within a cross-project or project premise. In estimating project defects, a model is constructed and used in a comparison. A lot of misinformation is needed for the project procedures. Therefore, the inter-project technique can be supported in a new research that requires more information for the programs. The inter-project disfiguration gauge (IPDP) is a way to implement one of the following assumptions by modifying models using innumerable data of a function: Studies in the field of IPDP have late. However, there are two shortcomings in the evaluation of pre-tests that cannot be emulated by the

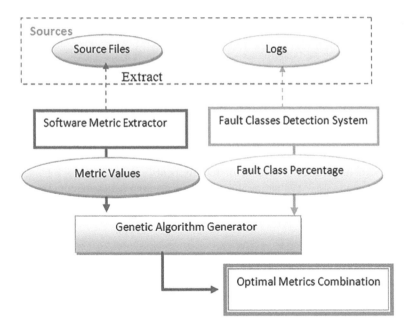

FIGURE 3.9   Genetic algorithm approach.

qualifications when applying the evaluation predictions or game planning strategies. Try to duplicate the IPDP procedures as suggested lately, and find the best way to do it based on the estimates. For example, for the F score, AUC and MCC results, a 7- or 8-year approach might be better. The approval of the request is reissued before it can be determined that it is too large. Riot and clean data were used, and comparative results were obtained from two sets of data. However, some aggregation estimates gave better results. The manufacturers later decided that the decision of the collection method had an effect on the presentation of the design. Different default assumptions are incorporated into DM techniques. When you go with the branches, we will explain these tests in the context where they practise learning the costume. The ES examines some of the flawed assumptions. Weaknesses using systematic learning methods combine group learning with two basic learning styles to achieve modelling rather than isolation. Different learning calculations; the differences of each comparison; available in a variety of configurations. In general, accumulation and promotion are explained in this section. Clicking (which solves the bootstrap assembly) is a form of equalisation. In this procedure, each form is created indefinitely and the data are sent to different subdivisions with different pre-delivery guarantees. It is therefore intended to reduce leisure. As a rule, it strengthens the majority of each social component. Promotion can be described as a progressive social event. First, the comparative loads are passed to the data events. In terms of preparation, the prevalence of misconceptions has increased and this association has on several occasions been the size of a meeting. Finally, this uses the popular weight loss program which means reducing the need. Stacking is a strategy that uses various concepts via a met classifier. Some of the distorting numbers are choosing the best course of action for a business: diffusion, promotion and abnormal trees. As you can imagine, groups such as forest choose the best driving experience (Rahman et al., 2019).

3.6 MACHINE LEARNING APPROACH FOR QUALITY ASSESSMENT AND PREDICTION IN LARGE SOFTWARE ORGANISATIONS REFERENCES

ISO defines quality as "the quality and characteristics of a product which affect its ability to meet communication or requirements". Requirements for use under "necessary conditions" evaluating the quality of programs that are consistent across the development cycle is the key to recognising and allocating the resources they need. Software forecasts provide a quantitative means of controlling programming objects and quality.

- Software quality assessment forms define the relationship between the required programming quality characteristics and quantitative skills.
- These models can be built with real strategies, for example backlit models or smart models.
- For example, because they are logical forms, neighbourhoods on decision trees or terms are white box models and prioritise their interpretation (Serban et al., 2020).

Programming Quality
- Software measurements have for some time been utilised for observing and controlling programming cycle, asses or potentially further develop programming quality.
- Metrics assortment and investigation is important for every day work exercises in huge programming improvement associations.
- Mature programming advancement associations likewise broadly utilise the data model of ISO/IEC standard 15939 as the method for distinguishing the data needs and executing estimation frameworks.

In this paper, we propose how Machine Learning-based methodologies can be utilised inside the ISO/IEC 15939 data model structure for successful appraisal and expectation of programming quality. The structure that utilises AI approaches inside the ISO/IEC 15939 data model will improve the reception of these strategies in enormous scope programming associations previously utilising the norm for their data needs (Harman, 2007).

3.6.1 ASSESSING SOFTWARE QUALITY ATTRIBUTES

3.6.1.1 Software Quality
With expanding significance of programming in our regular routines, the parts of value as for programming have additionally acquired high significance. Similar to

---

TABLE 3.2
Software Quality

Characteristics
Functionality
Reliability
Usability
Efficiency
Maintainability
Portability

Subcharacteristics
Appropriateness
Correctness
Interoperability
Safety
Functionality observance
Development
Mistake easiness
Recoverability
Dependability observance
Understandability
Learnability
Operability
Magnetism
Usability compliance
Time performance
Reserve competence utilisation
Effectiveness compliance
Analysability
Unpredictability
Permanence
Testability
Maintainability compliance
Flexibility
Installability
Coexistence
Replaceability
Portability

---

many aspects, the quality can be improved adequately in the event that we characterise it appropriately and measure it ceaselessly (Table 3.2). While quality is one of the
extremely normal and notable terms, yet it is equivocal and furthermore usually mis-
judged. To many individuals, quality is like what a government judge once said about
indecency "I know it when I see it". The primary explanations behind vagueness and
disarray can be credited to the way that quality is anything but a solitary thought,

yet a multidimensional idea, where measurements incorporate the element of premium, the perspective and the properties of that element. Along these lines, to completely see the value in the intricacies identified with quality the shift has been from characterising quality according to a solitary point of view towards characterising and working with quality models. Quality model as indicated by ISO/IEC 25000is: "characterized set of characteristics, and of connections between them, which gives a system to determining quality necessities and assessing quality" (Kim, 2020).

For definite portrayal of estimation data model and carrying out an estimation interaction, perusers are eluded to standard ISO. Two critical parts of the data model we would underline in this paper are furnished here with proper definition:

Estimation Parameter: The capacity is a calculation or calculation performed to join at least two base measures. It comprises balance- and scale-determined measure relying upon the scales and units of the base measures from which it is created just as how they are consolidated by the capacity.

(Investigation) Model: A calculation or computation consolidating at least one base and additionally inferred measures with related choice rules. It depends on a comprehension of, or suspicions about, the normal connection between the part measures as well as their conduct over the long run. Models produce estimates or assessments pertinent to characterised data needs. The scale and estimation strategy influence the decision of investigation methods or models used to create markers (Figure 3.10).

As clarified before, enormous mature programming advancement associations typically gather and screen different programming measurements considered significant with the end goal of screening and controlling programming improvement measure and delicate product/item quality. Given the accessibility of this enormous arrangement of information for current just as recorded undertakings and the unclarity of what low-level programming measurements mean for high request quality attributes (or generally speaking quality), we battle that for powerful appraisal and expectation of by and large programming quality in huge associations, AI strategies, for example, design acknowledgment and characterisation, can be utilised proficiently. In the structure, we first adopt a bottom-up strategy, considering that we have some quantitative evaluation of high request quality attributes (according to programming quality models, we can utilise machine understanding procedures for design acknowledgment such as CNN to perceive/anticipate under which quality class a given programming module/item falls at a given reason behind time during its turn of events). The model for such appraisal/expectation can be addressed as in Figure 3.10 (Rana and Staron, n.d.).

The model to assess the singular quality attributes can be acquired utilising hierarchical methodology as in ISO norms estimation data model. The following advances would be involved: first, relying upon the qualities of given delicate product project/item and requirements of various stake. Next, for distinguished data need (quality characteristics), subcharacteristics (comparing to inferred measures regarding ISO/IEC 15939) and various characteristics/programming measurements that might conceivably influence the given subcharacteristics are recognised. The subsequent

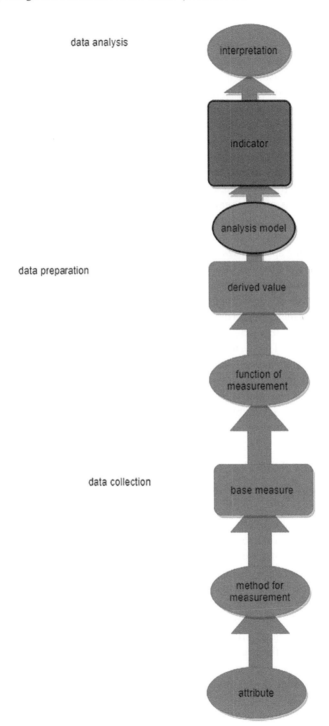

FIGURE 3.10 ISO/IEC 15939 measurement information.

stage is information assortment, which incorporates a collection of characteristics and utilises estimation hypothesis to allocate them esteems to acquire the applicable base measures. This progression likewise stays unaltered in our system regarding ISO/IEC 15939 measurement data model. Distinctive base measure(s) would now be able to be joined to frame inferred measures utilising design acknowledgment techniques (for example, fake neural networks) from the mama chine learning tool compartment. The fundamental benefit of utilising Machine Learning procedures in this progression is that utilising chronicled information, we can without much of a stretch and successfully utilise the example acknowledgment capacity of Machine Learning draws near, while discovering formal numerical relations for the equivalent is perplexing and troublesome. Subsequent to getting the quality subcharacteristics (inferred measures), we can again utilise the AI strategies, for example order models (for example, support vector machine) which can utilise the verifiable information to group given programming project/item/module to a class of value attributes. Again, Machine Learning apparatuses are profoundly helpful in this progression as tracking down the right examination model is troublesome and complex. The acquired quality attributes for current delicate product project/item/module would then be able to be deciphered. While AI approaches have been applied to numerous computer programming issues and furthermore to numerous singular programming quality attributes/subcharacteristics, generally their utilisation for quality appraisal and forecast is uncommon. The system introduced in this paper should be approved in an enormous programming association setting which we see as our future work bearing. We likewise accept that more examination is required around here to set up models for assessing and anticipating higher request quality attributes and in general quality utilising broadly accessible programming measurements information utilising AI procedures (Nascimento et al., 2020) (Figure 3.11).

3.6.2 Quality Prediction Using Threshold Euclidean Distance Model

The boundaries picked for the model depended on after suppositions.

- The mental separation needed to plan and execute a program relies on the quantities of strategies and number of variable names.
- The last lines of code created influence the advancement time.
- The sequence of strategies is an indicator of how much exertion is needed to foster a program.
- The programming language openness/experience of a software engineer influences the improvement time.
- The innate program trouble level (as experienced by the software engineers) additionally influences the advancement time (Rashid, 2012).

Metric Thresholds: In this review, the data gathered from understudies incorporated the following:

- No. of lines in code
- No. of functions used

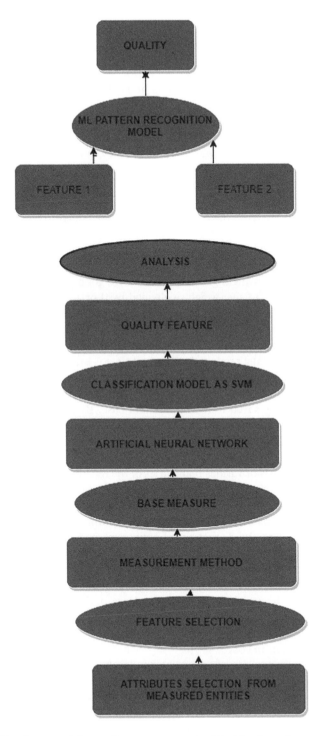

FIGURE 3.11 Framework for quality assessment using Machine Learning.

- Level of difficulty
- Knowledge scale metrics of programmers.

Comparability Function Used

Euclidean distance: This comparison is probably the most commonly used distance between highlight vectors. The distance is given by the client by taking the weight for each autonomous variable. Suppose that the journal $S1$ of n blocks contains the following attributes for each of the boxes $w1$, $w2$.... and n. Field # 1, $x2$. A similar log was recorded with xn. In this model, the Central Commission of Inquiry is another means by which the provisions relating to minorities to some extent affect the proximity of the project. Different approaches are proposed:

- Compose all undertaking highlight weight to indistinguishable qualities: w $o...w = 1$.
- Arrange each undertaking highlight weight to a worth controlled by human judgment.
- Arrange each task highlight weight to a worth got by factual examination.
- Accumulate each value weight is divided by 0 or 1. Improve the measure of evaluation quality. This powerful approach seeks to distinguish between separate provisions. When these parts are separated, they all give the same weight. Based on the information, the best competition is found based on the information and the product development time is expected. We have shown a comparison of the product quality with the value sample is extracted from the knowledge base ($q1$); assuming the error is less than 10%, the dataset is automatically saved to the information base. By noting incorrectly, the probability of a point is further investigated. The graphic of the proposed framework is shown in Figure 3.12. This creates an ace information base from a set of records (records).
- The given values of various boundaries of the record set are acknowledged.
- The difference of the info set is determined from each record set in information base.
- The difference is determined utilising few similitude parameters. For this situation, two similitude measures for given method are utilised.
- The ledger set(s) with least distance are the coordinating case(s).
- The anticipated improvement scale is advancement season of the coordinating with case. The framework predicts the nature of the product subsequent to tolerating the upsides of specific boundaries of the product. The boundaries include the following (Xing et al., n.d.):
  - Proportion of factors.
  - No. of lines in code.
  - Various types of method; complexity level.
  - Calibre of developers.
  - Expectation depends on relationship and condition thinking that utilisation different comparability measures.
  - If any instrument to refresh the information base (information base of cases) as new cases are created.

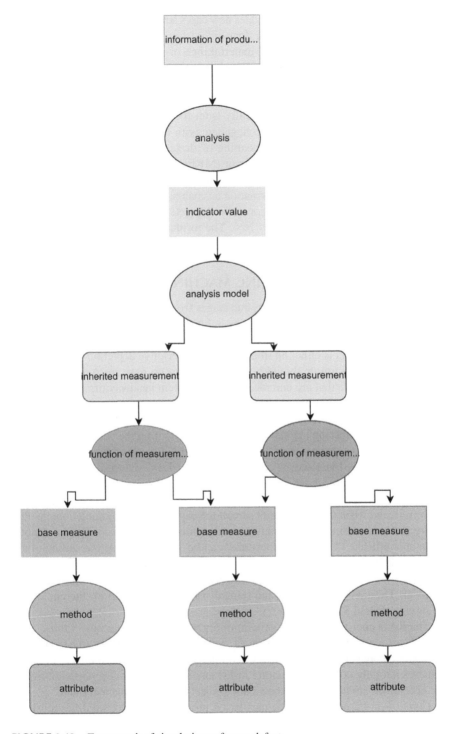

FIGURE 3.12 Framework of simulating software defect.

- The framework acknowledges new cases straightforwardly. Change of a specific record is likewise done.
- A director plays out the undertakings of refreshing and altering the data set.

Based on both the Manhattan distance and the Euclidean distance, the distance of each model is considered individually from the threshold vector. The real-time progress and prudence/proximity to negotiate with the case are extracted from the information base. Determine the meaning of these qualities and display them on the screen using the Yield Estimate Error (MRE). We have shown a comparison of the product quality with the LOC extracted from the knowledge base ($q1$). For this situation, we used 70% of the information as change data and 90% as test data. The estimate was 95.4% of the 10% error in the correction, and the prediction was 94% of the error for the test, which was 76%. The results are generally very good when practising case-based thinking (Figure 3.13).

3.7 MODEL SELECTION USING MACHINE LEARNING

An item cycle oversees various parts and stages from needing to testing and passing on programming. This heap of activities is finished startlingly, as per the necessities. Each way is known as a Software Development LifeCycle (SDLC) model. An item life cycle model is either an illustrative or prescriptive depiction of how writing computer programs are or should be made. Coming up next are some well-known fundamental models that are embraced by various item headway firms.

Waterfall approach: When essentials are obvious and stable, the course model also called the conventional life cycle, with its productive and progressive procedure, can be utilised. The gathering begins with correspondence from the customer concerning specific and advances through orchestrating, showing, improvement and association. If the essentials are fixed and expecting work proceeds in an immediate style to complete the endeavour, the course model is fitting (Technology and Road, 2016).

Prototyping approach: When clear essentials for lackness and features can't be perceived, and when the creator isn't sure of the capability of a computation, the flexibility of a functioning system, and the kind of human-machine affiliation, a model thought is utilised. "Used as a procedure" can be executed inside the setting of any of the cycle models. The model is made in the wake of fixing the overall objectives and essentials. The accompanying quick arrangement tops in the improvement of a model. The model is checked and refined with the analysis from the end clients (Hanselmann and Sarishvili, 2007).

The RAD approach is a slow programming headway measure model those anxieties an astoundingly short improvement cycle. The RAD model is a fast variety of the course model. The quick headway is refined through part-based turn of events. It achieves a totally utilitarian structure inside an outstandingly short period of time if the necessities are doubtlessly known and project expansion is constrained.

Component-based model: The part-based progression model solidifies a critical number of the characteristics of the winding model. It is formative in nature. It uses

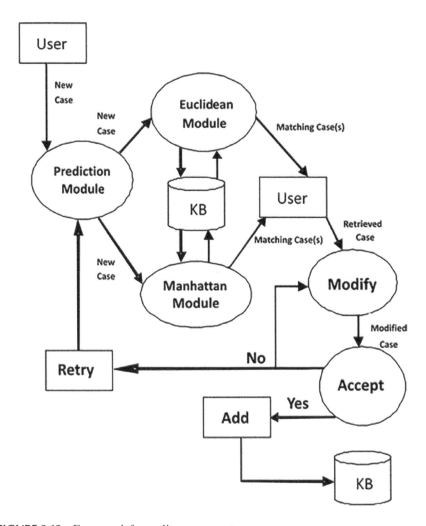

FIGURE 3.13　Framework for quality assessment.

existing reusable portions. The accentuation is on consolidating the parts as opposed to making them from the start. The endeavour cost and progression measure span can be diminished by joining section reuse as a component of the definitive culture. The portion-based model has various advances going from necessities assurance, part examination, essential change, structure plan with reuse, improvement and coordination and system endorsement (Wang et al., 2007).

3.7.1　Choosing a **SDLC** Model

Choosing the SDLC approach can be a daunting task for some relationships. Unique programming development lifecycle models are intertwined; conditions force dangers expense plan; suitable for work that sums up service life, etc. However, to meet the requirements, you must determine which model to compromise on. There are many ways to make a successful business venture. Some of them are inexperienced and depends on performance and customer needs. Approach the scores of efforts to recognise the qualities of each cycle model that can help us in adopting the article development model, which is a rough number approach. The winning method with the highest score here is a suitable plan of characteristics (Kapur and Sodhi, 2019):

- Are the prerequisites grounded, oral-characterised? Interface?
- Are the necessities determined or liable to change as the undertaking advances?
- Is the undertaking little to medium-evaluated (up to four individuals for a very long time) or large?
- Is the application like undertakings that the engineers have insight into, or is it another region?
- Is the product liable to be is it direct or complex (for example, does it utilise new equipment)?
- Does the product have a little simple UI or a huge complex client?
- Must all the usefulness be conveyed without a moment's delay or would it be able to be conveyed as incomplete items?
- Is the item security basic or not?
- Are the engineers generally unpractised or chiefly experienced?
- Does the hierarchical culture advance individual imagination and duty or does it depend on clear standards and methodology (Bhavsar et al., 2020)?

3.8   RESULTS AND DISCUSSION

For three types of data: 1; the accuracy of classifications 2 and 3 is shown in Table 3.3. As shown in Table 3.4, all three Machine Learning estimates achieved high precision.

TABLE 3.3
Model Accuracy and Score

Trend	Score	Model and Algo. Used
Requirements clarity	99%	SVM
Requirements change	67%	Random forest
Project size	75%	Gradient boost
Application	87%	TDF
Software	65%	IDF
User interface	87%	SVM
Functionality	89%	SVM
Safety critical	91%	Regression
Developer expertise	89%	Neural network
User involvement	93%	Decision tree
Total Score	**88.8%**	

TABLE 3.4

Accuracy Analysis

Dataset	NB	DT	ANNs
Sample 1	0.669	0.877	0.654
Sample 2	0.887	0.865	0.876
Sample 3	0.778	0.888	0.876
Average	**0.887**	**0.897**	**0.985**

For all three classifiers, the normal motivation for the accuracy rate takes everything into account. However, the lower value for the NB estimate appears in the DS1 dataset. We recognise that this dataset contains nothing and that the NB calculation requires a more memorisable dataset to obtain more precise data (Table 3.5). In this way, the NB achieved higher accuracy rates in the DS2 and DS3 datasets, which were significantly more pronounced than the DS1 datasets (Pradhan et al., 2020).

TABLE 3.5

Characterising Features of Project

Project Feature	Commentary as per Behaviour	Scaling
Supplies clearness	Settled	10
Necessities modified	Permanent	9
Venture dimension	Medium to average	8
Submission	Known	7
Application	Uncomplicated	7
UI	Undemanding	8
Toggle feature	Only once	5
Safety critical	No	3
Developer knowledge	Simple	8
Workforce	Independence of module	7
Consumer participation	Simple	6
Project attribute	Score is 1	5
Requirements clearness	Completed	5
Requirements contrast	Permanent	6
Scheme measurement	Small to medium	7
Submission	Wellknown	9
Software	Simple	6
Background	Simple	9
Uses	One time	6
Critical phase	No	9
Skill set measure	Principallyinexperienced	6
Backend	Autonomy	8
User contribution	Smallest	Broad

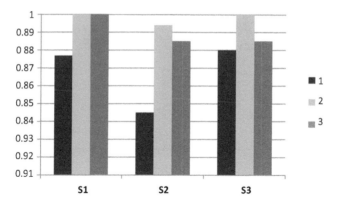

FIGURE 3.14   Accuracy comparison.

Table 3.4 provides specific measures for the classification of NB, DT and ANN in DS1, DS2 and the data. As a result, three Machine Learning calculations can be used for satisfactory assumptions at a fair rate. The overall accuracy of the three data classes is greater than 97%. The third measure is the audit measure. Table 3.4 shows the survey for three subtypes in three data sets. Also, this Machine Learning estimate is passed in terms of fair audit. The best survey is compiled by the DT classifier, which is 100% accurate in all data. On the other hand, the regular surveys for the ANN and NB calculations are almost completely independent by 96%. The *F*-measure is used to balance the three classifiers according to audit and specific procedures. The *F*-measure is respected for Machine Learning calculations used in three datasets. As Figure 3.14 shows, DT is an ANN; next to the NB classifiers is the most distinctive *F*-measure of all the data. Finally, the RMSE issue is not really posed to evaluate the Machine Learning estimate. A simple automatic regression (AR) model is provided to predict the total number of program problems using chronic evaluation errors. In the RMSE measurement, the POWM model area and their system were studied. Evaluation collaborative work is being done on the comparative data that we use in this evaluation (Chigurupati et al., 2020).

3.9   CONCLUSIONS

The main purpose of this work is to present assessment forms based on Machine Learning strategies. The submissions of these models vary depending on the quality given. In any case, it is essential that the data from programming projects and the integrated black box techniques used by SVM, in particular, provide consistent and unbelievably high accuracy (Laradji et al., 2015). Their ability to obtain high-speed data can be used to coordinate future glassware designs and future program developments. An essential advantage of these Machine Learning styles is that they are equal. We can combine them in a changing cycle like a robot. In any case, these forms can be used to access data from additional programs; it should be downloaded from a variety of applications. It is critical. This is an experiment to create reusable models (Nam, 2014).

Programming quality is the level at which communication or compliance with certain requirements and concerns is met (Wang and Yao, 2013). An item metric is a measure of an asset that does not degenerate into a system or a combination of items. Components are deformed by software during the early detection of faults using a subsequent identification method (Lincke et al., 2010). In this chapter, we have discussed different application methods used on a variety of datasets, based on the existing programming estimates. Going forward, we will look at the results of different methods of data collection and management game plans. Good quality programming will be supported. Quality is the key to writing computer programs. It is a term that is not yet available (Guo and Lyu, 2000). Understanding the factors that can affect quality and general standards helped us; the relevance and size of the impact of written components/subtypes on general attributes are grim. Features externally and later, it can be damaged by a large number of unnecessary components (Harman, 2007). We provide a framework that uses ISO/IEC 15939 assessment information and standardised AI strategies to effectively implement internal data to demonstrate program quality by using standards to fully demonstrate this information (Harman, 2007). The use of Machine Learning methods does not motivate the identification of specific groups in atmospheric measurements and requires careful consideration. Special sub-credits are defined as higher quality of demand or discipline. With the use of chronic data, Machine Learning techniques can help assess regulatory quality and high-demand quality registrations based on assessment criteria (Wang et al., 2007). Another essential advantage of using Machine Learning techniques is that they are used in these giant relationships. Their power of precision and fairness accumulates over time. This makes them incredibly cute for such an assessment. Standard programmer development models through SDLC can support the current situation (Bhavsar et al., 2020). Some models having influence and demand are guaranteed by engineers for their warranty. The SDLC is difficult to select if it is demonstrated that the effort illustrates the features of the programming development model (Pradhan et al., 2020). Conceptual application diagram confirms the authenticity of the chosen programming development model. In integration, the survey involves a simple framework and scores of semantic and material engineers, according to the scores, according to the programming characteristics (Sinha et al., 2020).

REFERENCES

Ahmed, M. R., Zamal, M. F. Bin, Ali, M.A., Shamrat, F. M. J. M., & Ahmed, N. (2020). The impact of software fault prediction in real-world application: An automated approach for software engineering. *ACM International Conference Proceeding Series*, 247–251. https://doi.org/10.1145/3379247.3379278.

Amershi, S., Begel, A., Bird, C., DeLine, R., Gall, H., Kamar, E., Nagappan, N., Nushi, B., & Zimmermann, T. (2019). Software engineering for Machine Learning applications. *ICSE*, 2020, 1–10. https://fontysblogt.nl/software-engineering-for-machine-learning-applications/.

Arora, I., Tetarwal, V., & Saha, A. (2015). Open issues in software defect prediction. *Procedia Computer Science*, *46* (ICICT 2014), 906–912. https://doi.org/10.1016/j.procs.2015.02.161.

Azar, D., & Vybihal, J. (2011). An ant colony optimization algorithm to improve software quality prediction models: Case of class stability. April. https://doi.org/10.1016/j.infsof.2010.11.013.

Bhavsar, K., Shah, V., & Gopalan, S. (2020). Scrumbanfall: An agile integration of scrum and kanban with waterfall in software engineering. *International Journal of Innovative Technology and Exploring Engineering*, 9(4), 2075–2084.https://doi.org/10.35940/ijitee.d1437.029420.

Challagulla, V. U. B., Bastani, F. B., Yen, I. L., & Paul, R. A. (2005). Empirical assessment of Machine Learning based software defect prediction techniques. *Proceedings – International Workshop on Object-Oriented Real-Time Dependable Systems*, 263–270. https://doi.org/10.1109/WORDS.2005.32.

Chigurupati, T. R. M., Basha, S. K. M., & Sunil, E. (2020). Constructive heart disease prophecy with hybrid Machine Learning strategy. *International Journal of Scientific Research in Computer Science Applications and Management Studies*, 9(3).

Chug, A., & Malhotra, R. (2016). Benchmarking framework for maintainability prediction of open source software using object oriented metrics. *International Journal of Innovative Computing, Information and Control*, 12(2), 615–634.

Dam, H. K., Pham, T., Ng, S. W., Tran, T., Grundy, J., Ghose, A., Kim, T., & Kim, C.-J. (2018). A deep tree-based model for software defect prediction. *Software Engineering*. http://arxiv.org/abs/1802.00921.

Feldt, R., De Oliveira Neto, F. G., & Torkar, R. (2018). Ways of applying artificial intelligence in software engineering. *Proceedings – International Conference on Software Engineering*, 35–41. https://doi.org/10.1145/3194104.3194109.

Fenton, N. E., Society, I. C., Neil, M., & Society, I. C. (1999). *Defects_Prediction_Preprint105579*, 25(3), 1–15. http://www.eecs.qmul.ac.uk/~norman/papers/defects_prediction_preprint105579.pdf.

Gray, D., Bowes, D., Davey, N., Sun, Y., & Christianson, B. (2011). The misuse of the NASA Metrics Data Program data sets for automated software defect prediction. *IET Seminar Digest*, 2011(1), 96–103. https://doi.org/10.1049/ic.2011.0012.

Guo, P., & Lyu, M. R. (2000). Software quality prediction using mixture models with EM algorithm. *Proceedings – 1st Asia-Pacific Conference on Quality Software, APAQS*, 2000, 69–78. https://doi.org/10.1109/APAQ.2000.883780.

Hammouri, A., Hammad, M., Alnabhan, M., & Alsarayrah, F. (2018). Software Bug Prediction using Machine Learning approach. *International Journal of Advanced Computer Science and Applications*, 9(2), 78–83. https://doi.org/10.14569/IJACSA.2018.090212.

Hanselmann, G., & Sarishvili, A. (2007). Heterogeneous redundancy in software quality prediction using a hybrid Bayesian approach. Berichte des Fraunhofer ITWM 125.

Harman, M. (2007). The current state and future of search based software engineering. *FoSE 2007: Future of Software Engineering*, 342–357. https://doi.org/10.1109/FOSE.2007.29

He, P., Li, B., Liu, X., Chen, J., & Ma, Y. (2015). An empirical study on software defect prediction with a simplified metric set. *Information and Software Technology*, 59, 170–190. https://doi.org/10.1016/j.infsof.2014.11.006.

Kapur, R., & Sodhi, B. (2019). Towards a knowledge warehouse and expert system for the automation of SDLC tasks. *Proceedings – 2019 IEEE/ACM International Conference on Software and System Processes, ICSSP 2019, May*, 5–8. https://doi.org/10.1109/ICSSP.2019.00011.

Kim, M. (2020). Software engineering for data analytics. *IEEE Software*, 37(4), 36–42. https://doi.org/10.1109/MS.2020.2985775.

Laradji, I. H., Alshayeb, M., & Ghouti, L. (2015). Software defect prediction using ensemble learning on selected features. *Information and Software Technology*, 58(July), 388–402. https://doi.org/10.1016/j.infsof.2014.07.005.

Li, M., Zhang, H., Wu, R., & Zhou, Z. H. (2012). Sample-based software defect prediction with active and semi-supervised learning. *Automated Software Engineering*, *19*(2), 201–230. https://doi.org/10.1007/s10515-011-0092-1.

Lincke, R., Gutzmann, T., & Löwe, W. (2010). Software quality prediction models compared. *Proceedings – International Conference on Quality Software*, 82–91.https://doi.org/10.1109/QSIC.2010.9.

Ma, Y., Luo, G., Zeng, X., & Chen, A. (2012). Transfer learning for cross-company software defect prediction. *Information and Software Technology*, *54*(3), 248–256. https://doi.org/10.1016/j.infsof.2011.09.007.

Nam, J. (2014). *Survey on Software Defect Prediction*.Master Thesis. http://citeseerx.ist.psu.edu/viewdoc/summary?doi=10.1.1.722.3147.

Nascimento, E., Nguyen-Duc, A., Sundbø, I., & Conte, T. (2020). Software engineering for artificial intelligence and Machine Learning software: a systematic literature review. *Software Engineering*. http://arxiv.org/abs/2011.03751.

Okutan, A., & Yıldız, O. T. (2014). Software defect prediction using Bayesian networks. *Empirical Software Engineering*, 19(1), 154–181. https://doi.org/10.1007/s10664-012-9218-8.

Pelayo, L., & Dick, S. (2007). Applying novel resampling strategies to software defect prediction. *Annual Conference of the North American Fuzzy Information Processing Society – NAFIPS*, 69–72. https://doi.org/10.1109/NAFIPS.2007.383813.

Pradhan, S., Nanniyur, V., & Vissapragada, P. K. (2020). On the defect prediction for large scale software systems-from defect density to Machine Learning. *Proceedings – 2020 IEEE 20th International Conference on Software Quality, Reliability, and Security, QRS*, 2020, 374–381. https://doi.org/10.1109/QRS51102.2020.00056.

Rahman, M. S., Rivera, E., Khomh, F., Guéhéneuc, Y.-G., & Lehnert, B. (2019). Machine Learning software engineering in practice: an industrial case study. *Software Engineering*, 1–21. http://arxiv.org/abs/1906.07154.

Rana, R., & Staron, M. (n.d.). Machine Learning approach for quality assessment and prediction in large software organizations. *2015 6th IEEE International Conference on Software Engineering and Service Science (ICSESS)*. https://doi.org/10.1109/ICSESS.2015.7339243.

Rashid, E. (2012). A survey in the area of Machine Learning and its application for software quality prediction. *ACM SIGSOFT Software Engineering Notes*, 37(5), 1–7. https://doi.org/10.1145/2347696.2347709.

Rawat, M. S., & Dubey, S. K. (2012). Software defect prediction models for quality improvement: A literature study. *International Journal of Computer Science Issues*, 9(5), 288–296.

Rodriguez, D., Herraiz, I., Harrison, R., Dolado, J., & Riquelme, J. C. (2014). Preliminary comparison of techniques for dealing with imbalance in software defect prediction. *ACM International Conference Proceeding Series*. https://doi.org/10.1145/2601248.2601294.

Serban, A., Van Der Blom, K., Hoos, H., & Visser, J. (2020). Adoption and effects of software engineering best practices in Machine Learning. *International Symposium on Empirical Software Engineering and Measurement*.https://doi.org/10.1145/3382494.3410681.

Sinha, A., Singh, S., & Kashyap, D. (2020). Implication of data mining and Machine Learning in software engineering domain for software model, quality and defect prediction. *Journal of Emerging Technologies and Innovative Research (JETIR)*, 7(10).

Sun, Z., Song, Q., & Zhu, X. (2012). Using coding-based ensemble learning to improve software defect prediction. *IEEE Transactions on Systems, Man and Cybernetics Part C: Applications and Reviews*, 42(6), 1806–1817. https://doi.org/10.1109/TSMCC.2012.2226152.

Tantithamthavorn, C., McIntosh, S., Hassan, A. E., & Matsumoto, K. (2016). Comments on researcher bias: The use of Machine Learning in software defect prediction. *IEEE Transactions on Software Engineering*, 42(11), 1092–1094. https://doi.org/10.1109/TSE.2016.2553030.

Tejaswini, P. L. S., Varsha, K. S., Yasaswini, P., & Yalamanchili, S. (2019). Software defect prediction using Machine Learning techniques.*2020 4th International Conference on Trends in Electronics and Informatics (ICOEI)(48184)*, 2, 1053–1057.

Wahono, R. S. (2007). A systematic literature review of software defect prediction: Research trends, datasets, methods and frameworks.*Journal of Software Engineering*, 1(1), 1–16. https://doi.org/10.3923/jse.2007.1.12.

Wang, Q., Zhu, J., & Yu, B. (2007). Feature selection and clustering in software quality prediction. *EASE'07 Proceedings of the 11th International Conference on Evaluation and Assessment in Software Engineering*, 21–32. http://www.bcs.org/upload/pdf/ewic_ea07_paper3.pdf.

Wang, S., & Yao, X. (2013). Using class imbalance learning for software defect prediction.*IEEE Transactions on Reliability*, 62(2), 434–443. https://doi.org/10.1109/TR.2013.2259203.

Washizaki, H., Uchida, H., Khomh, F., & Guéhéneuc, Y. G. (2019). Studying software engineering patterns for designing Machine Learning systems. *Proceedings – 2019 10th International Workshop on Empirical Software Engineering in Practice, IWESEP 2019*, 49–54.https://doi.org/10.1109/IWESEP49350.2019.00017.

Xing, F., Guo, P., & Lyu, M. R. (n.d.). A novel method for early software quality prediction based on support vector machine. *16th IEEE International Symposium on Software Reliability Engineering (ISSRE'05)*. https://doi.org/10.1109/ISSRE.2005.6.

Zanutto, D., Lorenzini, E. C., Mantellato, R., Colombatti, G., & Sanchez-Torres, A. (2012). Orbital debris mitigation through deorbiting with passive electrodynamic drag. *Proceedings of the International Astronautical Congress, IAC*, 4, 2577–2585.

Zheng, J. (2010). Cost-sensitive boosting neural networks for software defect prediction. *Expert Systems with Applications*, 37(6), 4537–4543. https://doi.org/10.1016/j.eswa.2009.12.056.

4 Ambiguity Based on Working and Functionality in Deployed Software from Client Side in Prototype SDLC Model Scenario

Anurag Sinha
IGNOU

Kshitij Tandon
Jaypee University of Engineering and Technology

Shreyansh Keshri
Kalinga School of Management

Hassan Raza Mahmood
FAST NUCES Chiniot-Faisalabad Campus

CONTENTS

DOI: 10.1201/9780367816414-4

4.1   INTRODUCTION

Requirements engineering (RE) is a pattern of creation and refinement of a software requirements specification (SRS). It executes a huge occupation in programming improvement life cycle since SRS-made artefacts, for instance structure arrangement, coding and testing for the item headway and the achievement of programming project, are basically established on the idea of SRS documents. Consequently, SRS is basic in programming projects. SRS helps as a bond in the beginning of the improvement until the focal matter of significant worth control. Thus, at this stage the usage of SRS is applied with certain requirement engineering is generally written in ordinary language. Regardless, ordinary language is basically uncertain. Ambiguity infers a word can be unravelled in more than one significance. The four most ordinary sorts of ambiguity in SRS are (i) lexical, (ii) syntactic, (iii) semantic and (iv) lazy leaning. Lexical ambiguity exists when a word has somewhere around two likely ramifications. Syntactic vulnerability is generally called structure dubiousness and appears when a progression of words can be changed over into more than one unique way in light of uncertain etymological development. On the other hand, semantic ambiguity is a sentence, which can be changed over into more than one way inside its exceptional circumstance. Besides, rational vulnerability arises when a sentence doesn't express and the given setting is absent or missing the necessary information to clarify its importance (Gupta et al., 2019).

RE measure is a basic advance, since SRS superiority issues are essentially significant for various programming project spaces. Once in a while, SRS quality is straightforwardly appraised as the primary driver of calamities in programming improvement projects. IEEE standards give the characteristics of a decent SRS. The qualities comprise unambiguity, accuracy, modifiability, culmination, recognisability and positioning for significance, consistency, dependability and evidence. In any case, a complete, exact and consistent SRS requires a detailed examination to accomplish the precision level. An obvious examination issue in RE is settling uncertainty, where equivocalness can be characterised as "an assertion having more than one significance". Apparently no single wide, comprehensive and accurate meaning of equivocalness is written in the programming work. Each definition gives just a few sections and bits of the total definition by ignoring the remainder of the definition. In every way, it frames a total comprehension of the current meaning of uncertainty in software engineering. The IEEE-suggested preparation for software requirements stipulation says, "A SRS is unambiguous if, and just if, each necessity expressed in that has just a single translation". The issue with the IEEE description is that there is no unambiguous determination essentially on the grounds that for any particular, there is consistently somebody who comprehends it uniquely in contrast to another person, similarly as there are no sans bug programs. There are two significant wellsprings of equivocalness: correspondence blunders and missing data. Correspondence blunders happen because of articulation inadequacies and the absence of logical data between the writer and the peruser. Missing information can be a direct result of various reasons, for example human factor, nonappearance of insight and summarise module. Till date a huge part of the investigation work on SRS vulnerability has not been accustomed in a planned manner, consequently making researchers and experts put a solid effort to oblige and evaluate. To give a planned and coordinated point of view on the investigation into SRS obscurity, this outline depicts the current status of the strength of assessment work open in the field of SRS unclearness. The outline fuses logical arrangement of the middle thoughts and associations that together epitomise the SRS ambiguity field. This logical order is facilitated around two fundamental estimations, particulars and gadgets with which we endeavour to portray SRS ambiguity. While these huge estimations are by and large suitable for the fundamental spaces of programming improvement, we are enlivened by the composing of particular sub-estimations that are fundamental for the work in the field of SRS vulnerability. This material might pave the way for the usage of the SRS vulnerability technique in projects. Furthermore, it gives an aide as a plan that helps researchers focus on the most proper courses of action open for a particular dubiousness (Hayman Oo et al., 2018).

English language subtleties disguise their real significance behind obscure or ambiguous language. It was accepted that quite a bit of this was because of messiness and that essayists could really take care of business; however, for peruses of vague language, changing isn't an alternative. All the more significantly, uncertainty in some cases precisely passes a creator's expectation. Lawful writings are in some cases purposefully vague. Necessities engineers have since quite a while ago perceived that normal language is regularly vague. Settling ambiguities in source records for prerequisites stays a space of dynamic examination. Specifically, scientists have not zeroed in on distinguishing ambiguities in lawful writings that administer programming frameworks, which is basic since ambiguities in legitimate writings can neither be disregarded nor be handily eliminated. Numerous ways to deal with settling vagueness in programming necessities depend on disambiguation or expulsion of the equivocalness. These may essentially not be a possibility for programmers tending to vagueness in a lawful content. This chapter investigates the vagueness in a lawful content from the US medical services space regardless of whether programmers can really take care of business. The initial step for engineers building HITECH-managed frameworks is inspecting the content of the guideline and concentrate in prerequisites from it. Sadly, extricating programming prerequisites from guidelines is incredibly difficult. In any event, perusing and understanding these reports might be past the capacity of expert specialists. Recognising vague explanations and understanding why those assertions are questionable are basic abilities for necessities engineers perusing lawful writings. Indeed, even outside of the lawful area, a lot of undetected uncertainty is viewed as one of the five most significant explanations behind disappointment in prerequisites examination. As far as anyone is concerned, this chapter is quick to look at distinguishing proof and grouping of ambiguities in a legitimate content with the end goal of programming necessities examination (Mazza, 1989).

4.2 BACKGROUND

4.2.1 Customer Relationship Management Software

Dealing with the full degree of the client fuses two related destinations: one, to give the connection and the total of its client confronting workers with a solitary, complete perspective on each client at each touch point and across all channels, and two, to equip the client with a solitary, complete perspective on the affiliation and its broadly comprehensive channels. CRM is often insinuated as a facilitated exhibit. The CRM structure has been developed, especially after headway in network establishment, client/specialist enlisting and business information applications. This improvement drives associations to depend upon CRM systems for offering more inventive sorts of CRM is everything except another thought, yet it relies upon the latest headway in enormous business programming development. Similarly, associations use this plan to win the trust and the steadfastness of their customers. This works with usefulness in business. To achieve this, the CRM system needs to interface front and regulatory focus applications to stay aware of associations and build customer dedication. Moreover, CRM utilises ERP structures to achieve its goals. Upgrades in ICT and the web system (WWW) suggest that CRM structures could take advantage of these progressions with their ability to accumulate and look at the data on customer plans and translate customer direct. Additionally, associations can make a 360-degree viewpoint on customers to acquire from past associations with advance future ones (Eckerson and Watson, 2000). This advancement has conveyed one more importance to develop customer associations and proposes the new suggestion of "e-customers". The impact of ICT has been so remarkable that it effortlessly influences the overall advancing. Thusly, all affiliations have modified the business community. It is prominent that holding customers is more gainful than building new associations. Also, the progressions in CRM thoughts expect a critical part in additional fostering all items used in various associations, for instance, financial and movement business, and adaptable and vital associations. Furthermore, CRM techniques focus on the customers and compose requirements of the relationship around the customer rather than the thing. As shown by the above discussion, "Managing a viable CRM execution requires an organized and changed approach to manage advancement, connection, and people" (Alferoff and Knights, 2008).

FIGURE 4.1 CRM life cycle.

In Figure 4.1, an example study characterises CRM investigation into four primary classes: data frameworks and data innovation, advertising, deals, administration and backing. The greater part of the past distributions was unified on information frameworks (IS) and data innovation (IT). Along these lines, unmistakably IS and IT assume an extraordinary part in creating CRM. In any case, some open actions take various bearings, such as administration and client security.

The example study centres on CRM research from 2000 to 2005 and groups another plan of CRM into the accompanying fundamental classes: reception, obtaining, execution, use and support, development and retirement. It is qualified to specify that these stages were at that point used to portray the picture of big business asset arranging (ERP) framework. The disadvantage of examination is that it centres just around the diaries and gatherings for IS and promoting in Figure 4.2. Different trains such as administration, innovation, authoritative conduct and client conduct have been prohibited (Massey et al., 2014).

As the subjects of CRM research are hard to decide, the pertinent points through diaries and worldwide gatherings in IC and PC sciences or financial business sciences must be determined. The viewpoints of CRM are resolved in the accompanying subsection as per these two significant fields of CRM research (Figure 4.3).

Data Systems (IS) and Computer Science (CS) – Through the perception of CRM distributions, IS and its applications seem, by all accounts, to be a significant apparatus and significant point of view of CRM. In IS, CRM is the hidden foundation for

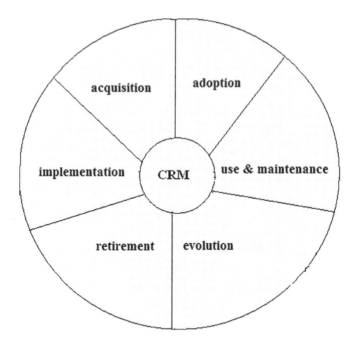

FIGURE 4.2   The life cycle of CRM.

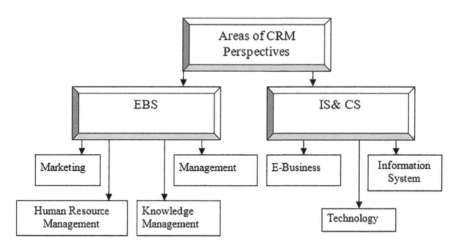

FIGURE 4.3   Viewpoint of CRM.

comprehension and connecting with clients effectively. The CRM points of view in IS and CS research are as follows:

Data System (IS): IS assumes a critical part in the advancement of CRM (Kincaid, 2003; Ling and Yen, 2001). The accentuation on IS discipline features on the significance of mechanical parts of CRM, a mix of programming, equipment and cycles and all applications lined up with client technique are clarified.

Innovation: The primary classes that portray this point of view are the modules of CRM, for example power computerisation, showcasing mechanisation, client care and backing. In addition, CRM as programming is given by numerous sellers in the business market.

E-Business: According to this point of view, CRM is a use of e-business and computerised exercises just as client assortment information.

The Business and Economic Science (EBS) – CRM is an "endeavour way to deal with comprehension and affecting client conduct through significant correspondences to further develop client obtaining, client maintenance, client dedication, and client productivity". From this perspective, CRM can be depicted by utilising the monetary and business points of view with their classes as follows:

The Board: It is the capacity frequently connected with CRM. CRM is established on showcasing and relationship advertising. Here, CRM frameworks are depicted as a business technique in excess of an innovation. The points of this viewpoint can be summed up in dealing with the client life cycle, expanding the devotion to the client, benefit and maintenance, which are the goals of the CRM framework.

Promoting: Most destinations that can be accomplished through this viewpoint can be exhibited in the accompanying focuses (Alferoff and Knights, 2008):

- Emphasis on long-term relationships and one-to-one cooperation through correspondence channels.
- Strong association among CRM and administration.
- Data gathered which are significant for special techniques.
- All types of administrative promotions.
- Some types of client administrations.
- Definition of CRM as utilisation of the CRM idea using ICT in both customary and electronic conditions.

Information Management (KM): In the information board, CRM implies learning the clients better to accomplish their destinations.

Human Resource Management (HRM): This point of view proposes the selection of a client situated culture by both top administration and representatives inside an association (Alokla et al., 2019).

4.2.1.1 Major Applications of CRM

Phone and Financial Credit Management: CRM programming helps bargains, displaying, and organisation specialists catch and track pertinent data about each past and orchestrated contact with conceivable outcomes and customers similarly as other business and life cycle events of customers. Information is obtained from all customer contact centres such as telephone, fax, email, the association's website, retail stores, stands and individual contact. CRM systems store the data in an ordinary customer informational index

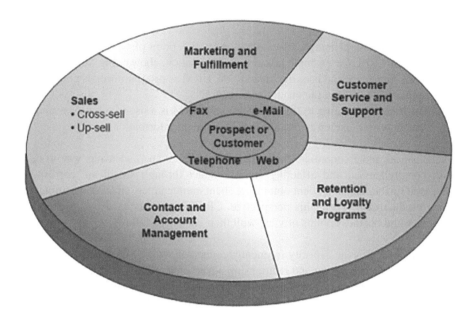

FIGURE 4.4 Application of CRM.

that arranges all customer account information and makes it available all through the organisation through Internet, interface or other association joins for bargains, displaying, organisation and other CRM applications as shown in Figure 4.4 (Alokla et al., 2019).

Deals: A CRM system gives sales representatives the item gadgets and companion's data sources they need to help and manage their arrangements and actuates and smooths out deliberate pitching and upselling. Deliberate pitching is a technique where a customer of one thing or organisation, say mishap inclusion, may similarly be keen on purchasing an associated thing or organisation, say contract holder's insurance. By using a deliberate pitching technique, specialists can all the more promptly serve their customers while simultaneously further fostering their deals. Up-offering suggests the way towards finding ways to deal with selling another or existing customer an ideal thing over they are correct now chasing. Extra models fuse bargain prospects and thing information, thing plan, and arrangements quote age capabilities. CRM in like manner gives steady across to a lone ordinary point of view on the customer, enabling sales representatives to watch out for all pieces of a customer's record status and history before arranging their business calls. For example, a CRM structure would alert bank specialists to call customers who set to the side enormous portions to sell them boss credit or theory organisations (Alokla et al., 2019).

Promoting and Fulfilment: CRM systems help advancing specialists accomplish direct publicising endeavours through automating such tasks as qualifying leads for assigned exhibiting, and booking and following direct mailings. Then, at that point, the CRM programming helps exhibiting specialists get and direct possibility and customer response data in the CRM informational collection, and separates the customer and business worth of an association's prompt publicising endeavours. CRM, moreover, helps in the fulfilment of prospect and customer responses and requests by quickly arranging bargain contacts and giving fitting information on things and organisations, while getting critical information for the CRM information collection.

Customer Care and Support: A CRM system offers support representatives with programming gadgets and progressing induction to the typical customer database shared by bargains and exhibiting specialists. CRM helps customer with changing bosses make, designate and administer requests for organisation by customers. Call centre programming reiteration calls to customer help experts subject to their capacities and ability to manage express kinds of organisation requests. In spite of the fact that language specialists see dubiousness or consensus as having a solitary, though wide, which means that is some of the time used to drive peruses to go to their own agreement or translation, unequivocally express that deficiency is a type of designing equivocalness that should be tended to for plan helps customer with changing representatives help customers who are having issues with a thing or organisation by offering material help data and thoughts for settling issues. Electronic self-organisation engages customers to get to altered help information adequately at the association website, while it's everything except a decision to get further assistance on the web or by phone from customer support workforce.

Retention and Loyalty Program: Improving and enhancing client maintenance and devotion is a significant business procedure and essential goal of client relationship the board. CRM frameworks attempt to assist an organisation with distinguishing, prize and market to their generally steadfast and beneficial clients. CRM scientific programming incorporates information mining apparatuses and other logical promoting programming, while CRM data sets may encompass of consumer information distribution centre and CRM information shops. These instruments are utilised to distinguish beneficial and steadfast clients and to coordinate and, what's more, assess an organisation's designated advertising and relationship showcasing programs towards them (Alokla et al., 2019).

4.2.2 Overview of SDLC and Prototype Model

Coordinated endeavour the leader's procedures (such as a SDLC) redesign the board's order over projects by parcelling complex tasks into sensible sections. An item life cycle model is either a realistic or prescriptive depiction of how writing computer programs is or should be made. Regardless, none of the SDLC models talk about the main issues of interest such as change the load up and incident organisation and release the leaders' measures inside the SDLC collaboration; simultaneously, it is tended to in the overall endeavour the chiefs. In the proposed theoretical model, the possibility of customer engineer association in the standard SDLC model has been changed over into a three-dimensional model that includes the customer, the owner and the architect. In the proposed hypothetical model, the possibility of customer creator correspondence in the customary SDLC model has been changed over into a three-dimensional model that incorporates the customer, the owner and the designer. The one-size-fits-all approach to manage applying SDLC systems is now not appropriate. We have made an undertaking to address the recently referenced acquiescence by using one more theoretical model for the SDLC depicted elsewhere. The disadvantage of watching out for these organisation measures under the overall endeavour the board is missing of key particular issues identifying with programming progression measure that is, these issues are talked in the undertaking the leaders at the surface level anyway not at the ground level (Baars, 2006).

Associations may utilise a SDLC model or elective system while dealing with any venture, including programming advancement, or equipment, programming, or administration obtaining projects. Despite the technique utilised, it ought to be custom-made to coordinate with an undertaking's qualities and dangers. Sheets, or board-assigned advisory groups, ought to officially endorse project strategies, and the executives ought to endorse and record critical deviations from supported techniques. Organised undertaking of the executives methods (such as a SDLC) upgrades the board's authority over projects by partitioning complex errands into reasonable segments. Sectioning projects into coherent control focuses (stages) permits chiefs to survey project stages for fruitful finishing prior to designating assets to resulting stages. The main stage in project management is where client require basic understanding of system being developed. A sequential process may simply join thoroughly portrayed stages, for instance prepare, acquire, test, do and stay aware of. Typical programming improvement projects consolidate beginning, organising, plan, progression, testing, execution and backing stages. A couple of affiliations consolidate a last, expulsion stage in their endeavour life cycles. The activities completed inside each adventure stage are similarly established on the board framework. All assignments should follow a lot of coordinated plans that clearly portray the essentials of each project stage. Accentuation further develops a project's ability to gainfully address the essentials of each get-together (end customers, security heads, originators, engineers, structure trained professionals, etc.) all through a project life cycle. Accentuation furthermore allows project bosses to complete, review and change stage practices until they produce adequate results (stage expectations) (Kamsties et al., 2001).

A project cycle oversees various parts and stages from expecting to testing and sending programming. This heap of activities is done surprisingly, as per the necessities. Each way is known as a SDLC model. A programming life cycle model is either a connection with or a prescriptive depiction of how writing computer programs is or should be made. An explaining model depicts the verifiable setting of how a particular programming structure is made. Clear models may be used as the justification behind understanding and further creating programming improvement measures or for building observationally grounded prescriptive models (Ezzini et al., 2021).

4.2.2.1 Prototyping Approach

The prototyping approach is a famous type of iterative SDLC that delivers a little model or form of the framework that the client can work with to give ideas. The

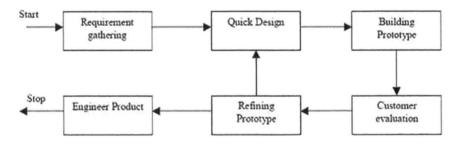

FIGURE 4.5   Prototype model.

methodology isn't an independent technique, but a way to deal with taking care of
bits of the bigger entirety. The ideas are then joined to make the framework completely functional. The prototyping approach is outlined in Figure 4.5. The figure
shows a circle through fast plan, constructing the model, client assessment and refining the model. This circle proceeds until the client is happy with the model, and
refinements have been carried out. By then, the model would then turn into the item.
The prototyping approach endeavours to decrease the danger by having the venture
in more modest pieces to ease changes required during the improvement stage. The
approach considers different emphases; be that as it may, the impediment happens
with numerous cycles. It is expected that the models will be disposed of and fruit-
less. This supposition that is incompletely because of knowing the prerequisites can
change definitely in the following cycle. For instance, the client could require another
element after a few models. The new element can change the extent of the issue
prompting degree creep. This prompts an exercise in futility and cash. Because of the
exercise in futility and cash, this methodology isn't reasonable for enormous scope
projects. Different models are additionally an administration calamity. The different
changes to fulfil the client not exclusively are hard to oversee yet additionally upset
the advancement group. The prototyping approach is best utilised for brief exhibits
or frameworks that have not been created. These sorts of frameworks can start the
establishment because of unsteadiness in another framework (Nacheva, 2017).

4.3   RELATED STUDY: PROTOTYPING MODEL
BASED ON PROCEDURE METHOD

The proposed model in Figure 4.6 is viewed as an iterative developmental prototyping measure that gets certain information sources, plays out a couple of steps and
conveys yield antiques. The current investigation offers the accompanying phases of
prototyping dependent on summed up strides of issues tackled in the writing audit:
framework prerequisites examination (compares with dissecting the issue), portraying (relates to fostering an arrangement), model turn of events (compares with carrying out the arrangement), investigating ease of use (relates to assessing results) and
refinement (Bano, 2016).

Information measure boundaries are the framework necessities and the picked
advances and instruments for programming improvement. They give the essential

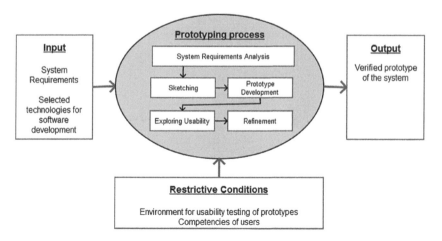

FIGURE 4.6   Prototyping model-based procedure.

premise to play out the interaction. As a yield antique of the prototyping cycle is made, a checked model that in the advancement interaction ought to be additionally improved with a view that the proposed approach depends on developmental prototyping, the prohibitive conditions for leading the cycle are related to the stage "Investigating Usability". Specifically, these are the capabilities of clients who will partake in the prototyping interaction and the climate where it will be led. The primary phase of the cycle is system requirements analysis. Its motivation is to attempt an appraisal of the principle communication situations with the framework according to the client's point of view. This needs to show the principle route streams, which requires recognisable proof of the fundamental entertainers in cooperation situations with the framework; the principle elements of the branch of knowledge and their various levelled association, assuming any, to make the underlying data model; the model is needed the basic difficult situations of algorithmic propagation, which are introduced according to the viewpoint of clients, as far as carrying out the interface associations. As a reason for defining the present data model, information design of the application or these are navigational components through which the client deals with the application (not really consolidated in a standard client menu); user interface components are engaged with the data stream, so their number could be somewhere in the range between 5 and 9; for example, these will be planned by the "7 ± 2" rule, giving some certainty that clients of the framework won't place pointless intellectual assets in working with interface (Bano, 2016).

4.4   LITERATURE REVIEW

Most programming prerequisites particulars are written in regular language, which is intrinsically questionable and uncertain. Be that as it may, programming designers don't yet have a solitary extensive and for the most part acknowledged meaning of ambiguity (Ferrari et al., 2014). Uncertainty is characterised as an expression that has more than one understanding (Dalpiaz et al., 2019). The training suggested by the

IEEE for programming prerequisites details expresses that explanation of necessities is unambiguous just when every prerequisite has a solitary understanding (Fantechi et al., 2018). Etymologists have adulated the meaning of equivocalness. In this part, we present the business related to the necessary arrangement and improvement. The presence of mind directs that an unambiguous assertion can just have one clear understanding. However, how would we characterise explanations that have no understanding? Ambiguous or fragmented articulations might not have legitimate understanding. To a necessities engineer, an assertion dependent on area information might appear to be befuddling from the beginning. Here, we believe dubious or deficient proclamations to be questionable on the grounds that they are not unambiguous. That is, we consider them questionable, on the grounds that they don't have a solitary and clear translation. Prerequisites specialists can permit numerous understandings of necessities right off the bat in the advancement of another arrangement of programming necessities (Gervasi et al., 2019). Moreover, a few assertions might be harmless on the grounds that only one potential change would be sensible, and such proclamations are probably not going to prompt misconceptions (Ganpatrao Sabale, 2012). Prerequisites for explanations with a few sensible understandings are unsafe and can prompt false impressions if not clarified (Yang et al., 2010). Rauterberg et al. (1995) spread the word about a further differentiation between perceived ambiguities for engineers and unnoticed ambiguities obscure to engineers. Many ways to deal with equivocalness in programming include the advancement of instruments or techniques to perceive or dispose of vagueness in programming prerequisites. For instance, Gordon and Bro use explanations to determine possible struggles between guidelines in various locales (Hammer and Vogel, 2013). The analysts utilised normal language handling to distinguish and kill equivocalness in programming necessities Abduljalil and Kang (2011) developed a way to deal with AI to recognise ambiguities in prerequisites. Chen and Popovich (2003) fostered a self-loader cycle to decrease the vagueness of programming prerequisites through object-arranged modelling. According to Grieskamp et al. (1998), ambiguities can be settled on the off chance that we know the setting of PE. Moreover, the creator considered the setting of not set in stone that an equivocal necessity is a prerequisite that has various implications. He depicted the significance of the ER setting, as practically all normal language prerequisites are probably going to be vague. Perusing the necessities, the greater part of the prerequisites can be dispensed with by the peruser who comprehends the setting of PE, and the remainder of the prerequisites we consider is vague. Phonetic vagueness (syntactic, lexical, semantic, over-simplification, vulnerability, and so on) doesn't rely upon any unique circumstance. The equivocalness explicit to RE relies upon the framework area, the program space, the advancement space, and the vagueness of the RE setting (Nacheva, 2017).

4.5 METHODICAL REVIEW: TYPES OF REQUIREMENT AMBIGUITIES AND THEIR DETECTION

4.5.1 Ambiguity in Requirements Engineering

The presence of mind proposes that an unambiguous assertion would have just a solitary, clear understanding. In any case, how could we order proclamations that have no understandings? Ambiguous or deficient articulations might not have a substantial translation. For a prerequisites engineer, an explanation that relies vigorously upon space information may likewise, from the outset, seem uninterruptable. In this, we believe dubious or fragmented proclamations to be vague since they are not unambiguous. That is, we believe them to be questionable in light of the fact that they don't have a solitary, clear understanding. Necessities architects may endure prerequisites with numerous understandings right off the bat in the advancement of another arrangement of programming necessities. What's more, a few assertions might be harmless in light of the fact that only one potential translation would be sensible, and these assertions are probably not going to prompt mistaken assumptions. Necessities with proclamations having more than one sensible understanding are toxic and liable to prompt misconceptions if not explained. Lawful area information would be needed to separate among harmless and poisonous necessities in this examination. Since we don't accept our contextual analysis members have the essential foundation, we don't consider the distinction among toxic and harmless to be significant. It makes an extra qualification between recognised, which are known to engineers, and unacknowledged ambiguities, which are obscure to engineers. Numerous programming ways to deal with uncertainty include the improvement of devices or methods for perceiving or disposing of equivocalness in programming prerequisites. For instance, analysts have utilised regular language handling to identify and resolve vagueness in programming prerequisites. In spite of the fact that contentions between strategy records, legitimate writings and programming prerequisites may not really be a type of uncertainty, these struggles enlivened our work in two essential manners. To begin with, it expresses that arrangement among strategies and programming necessities must be impeccable to keep away from clashes. Indeed, even potential struggle ought to be tended to. These statements support the utilisation of a wide meaning of equivocalness (Pittke et al., 2016).

4.5.2 Types of Ambiguity

Lexical Ambiguity: This happens when an expression or articulation has various genuine ramifications. Consider § 170.302(d): "Enable a customer to electronically record, change, and recuperate a patient's powerful medication list similarly as medication history for longitudinal thought". A remedy history for longitudinal thought could mean either an absolute solution history in a particular strategy or a compressed medication history used extraordinarily for a particular explanation. A necessities specialist ought to disambiguate this before execution. Another model: "Melissa walked around the bank". This could infer that Melissa walked around a financial foundation or she walked around the edge of a stream as shown in Figure 4.7.

Syntactic Ambiguity: This happens at what time a progression of words has various genuine phonetic parsings. Consider "Engage a customer to electronically record, adjust, and recuperate a patient's fundamental signs...". Here, "electronically" may imply all of the activity words "record, change, and recuperate" or just to "record". It seems, by all accounts, to be inconceivable that the US government needs EHR shippers to "electronically

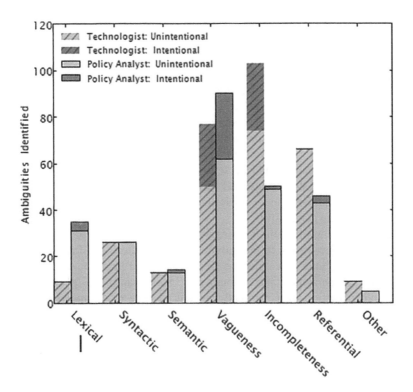

FIGURE 4.7   Ambiguity type.

change a patient's basic signs". But, electronic record or recuperating seems like reasonable prerequisites. Again, a necessities specialist ought to disambiguate this before execution. As well: "Quickly examine and talk about this entry" (Bäumer and Geierhos, 2018).

Semantic Ambiguity: A sentence occurs when there is more than one meaning, depending on the envelope specification. Each word in a sentence has an unmistakable meaning, and the sentence has a separate tree; however, the correct understanding of the sentence should be emphasised. Consider 170.302 (j). If a parameter is defined for communication, it is reasonable to compare the two records. This is the only time. Cost quantities can be calculated for drug-related or multiple components. Likewise, these summaries may have room for different or different patients depending on the stimulus of the relationship.

Unclearness: This happens while a phrase or proclamation concedes marginal belongings or comparative understanding. Consider "Electronically quality, partner, or connection a research facility test result to a lab request or patient record". What establishes ascribing, partner or connecting? Must these records consistently be shown together or would essentially having an identifier and permitting a doctor to discover one given the other do the trick? Additionally, consider: "Fred is tall". If Fred was a North American male and 5′2″ tall, then at that point the case isn't correct. In the event that

Fred was 7'0" tall, the case is upheld. Some place in the middle of falsehood statures that sensible individuals may differ as to comprising "tall" (Pittke et al., 2016).

Deficiency: This happens when an assertion neglects to give sufficient data to have a solitary, clear understanding. Consider § 170.302(a)(2): "Give certain clients the capacity to change warnings accommodated drug-medication and medication sensitivity cooperation checks". This sentence precludes data that would permit prerequisites specialists to distinguish which clients ought to have this capacity for sure alternatives they would need to change notices. Deficiency should be settled for the necessities to be executed. Likewise, "Join flour, eggs, and salt to make new pasta". This precludes some important data, for example amount of materials and procedures to be utilised (Yang et al., 2010).

Reference Ambiguity: This happens while an expression in a judgement can't be said to encompass an instantly recognisable reference (Table 4.1). "For each significant utilise objective with a rate based measure, electronically record the numerator and denominator". Significant usage goals that use baseline metrics are not directly referenced. The requirements to calculate which sites should comply with these legal commitments are omitted. Different forms include pronouns and their prepositions. "Taught the child to his father about the vulnerability can be referred to. He (the father). In addition, the" legal counsel to lie for many reasons. Some are better than others in the administration. "Depending on the types of uncertainties relating to the stability of our countless scientific classifications of people. Being fairly broad and will be thorough about the unexpected.

TABLE 4.1
Ambiguity Description

Ambiguity type		
Definition		**Example**
Lexical	A sound or expression with numerous convincing meanings	Melissa amble to the depository.
Syntactic	A succession of words with several applicable grammatical explanation in spite of context	Quickly interpret and talk about this tutorial.
Semantic	A verdict with more than one elucidation in its provided circumstance	A and B are married.
Vagueness	An account that acknowledges norm cases or comparative explanation	Fred is big.
Incompleteness	A grammatically accurate judgement to facilitate provides moreover modest element to suggest a detailed or needed significance	Merge flour, eggs, and brackish to construct fresh pasta.
Referential	A grammatically truthful judgement with a location that confuse the person who reads based on the framework	The schoolboy told his minister on the subject of the smash-up. He was awfully saddened.

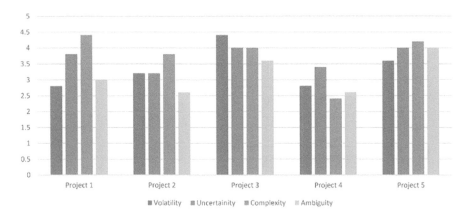

FIGURE 4.8 Ambiguity measurement per project.

Wording, meaning more than one word, a sentence or a paragraph is not compatible with our scientific classifications. This related content for the analysis, such as other members of uncertainty. Note command. HITECH Act as it has been sentenced to content 0 in the progression of the program because it was a part of our collection. For its obvious consequences, this investigation is not intended to ensure that suspicion arises on a large scale. It can also be assumed that a paragraph in the text contains only one translation that has a clear meaning. In our scientific category, these explanations are called unambiguous articulations (Dalpiaz et al., 2019; Figure 4.8).

4.5.3 Approach of Literature Segmentation for Resolution of Ambiguity Detection

Based on removed writing, we determined the classification to get the essential thoughts and connections free from SRS equivocalness. This scientific categorisation is ready around two key measurements – specialised measurement and devices measurement, as displayed in Figure 4.9. With them, we portrayed the insights into SRS equivocalness. We utilised an appropriate determination technique for each key measurement, as a predefined objective to get significant papers from the huge arrangement of writing. Nonetheless, the required specialised instruments and measurements are not restrictive, these measurements are connected to the different issues of SRS, and here we will zero in just on one issue for example equivocalness. We discovered writing explicit to sub-measurements that are critical and of good importance in setting to determine the equivocalness (Mich and Garigliano, 2000).

Here, we are clarifying momentarily the two measurements and their connected significant targets. The significant point of the specialised measurement is in the direction of portraying the kinds of uncertainty, strategies, procedures and representation to determine the vagueness. On the way to accomplish this, every advance should be portrayed with precondition life cycle including various stages and now and again it very well may be a mind-boggling measure. The idea of the interaction

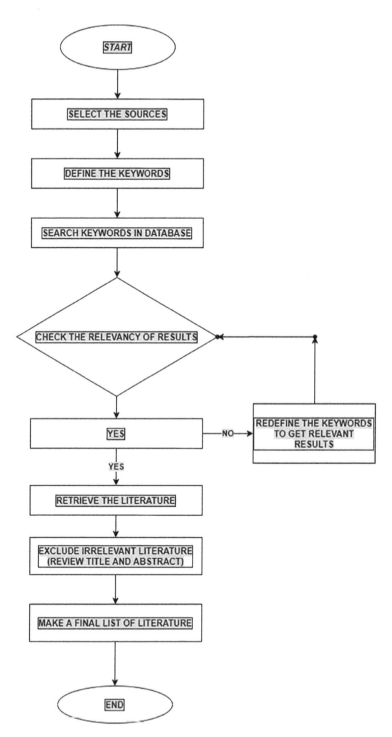

FIGURE 4.9   Flowchart of literature classification.

relies upon the representation to decide for your software advancements such as V, Double V, Waterfall and Incremental. This studied the writing and discovered different issues in picking the prerequisite model. There are reasons that can uncommonly impact the ambiguities, the assorted gathering occupations (bunch occupations, size of the attempt and assurance) and the methodology applied to recognise and take out obscurity in SRS reports. Finally, the device estimation portrays how instruments can support to perceive and dispose of SRS issues. For this estimation, we portrayed the work of the various gadgets (reason) and investigated what system is used by the different gadgets as there are a great deal of contraptions that work on different sorts of ambiguities. More than two estimations may not be absolutely specific as a piece of the limits may be covered, concerning now, this is unavoidable. For example, strategies discussed in the development might be used as a base for the gadget in the other estimation (Figure 4.10). Here, instruments are simply inferred for the modified distinguishing proof and departure of the ambiguities so we can reduce the overall expense and can save important period of the gathering. However, we put forth a legit attempt to all the more probable clarify these estimations from the end customer perspective. By and by, we will pursue our discussion on the two estimations and their

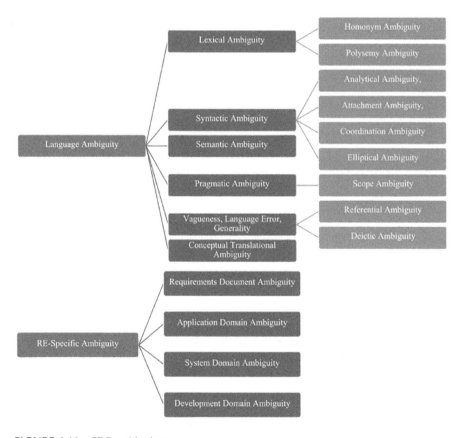

FIGURE 4.10   SRS ambiguity.

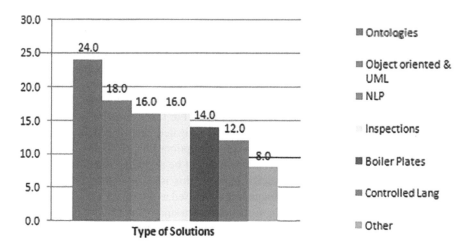

FIGURE 4.11   Solution for ambiguity detection.

sub-estimations with the help of appropriate material which we removed from the databases by using related expressions (Bäumer and Geierhos, 2018).

Here, our principle centre is to recognise existing arrangements which can work on the nature of the SRS as far as equivocalness and location decrease. We explored and examined 54 writings and found the greater part of the arrangements can be isolated into six principle classifications dependent on the method utilised by specialists to determine vagueness. In the overview, we found that answers for SRS vagueness can be extensively arranged into six classifications (Chen and Popovich, 2003; Figure 4.11):

  a. Ontology-oriented answer
  b. OOPs-based foundation objective
  c. Natural language understanding-formed answer
  d. Examination support result
  e. Algorithmic explanation
  f. Unusual idiom support explanation.

4.6   METHODOLOGY

4.6.1   Data Collection and Survey

We have done a methodical survey to gather responses of industrialist for preference and having flexible documentation of SRS and in working functionality of software.

4.6.2   Proposed Model

4.6.2.1   Enhanced Prototype Model

We recommend an adjusted prototyping model, which is a changed rendition of the past work. In contrast to the past work, it has not been misused and tended to clients' input seriously similar to our projected model. The disadvantage of the past work may for the most part postpone the engineers at certain means of the plan because of the absence of clients' inputs. Some of computer programmers may ask why we need to remember the client criticism for each phase of the product plan. It is basically on the grounds that clients need a framework that is liberated from blunder and simple to cooperate. To do so, we need a client to take care of us with assessments and remarks to stay away from any undesirable and bothersome highlights. The accompanying advances clarify the design of our model in Figure 4.12 (Nacheva, 2017; Table 4.2).

In order to fully capture an individual's emotional behaviour when interacting with apps, we need things and regulations that allow us to collaborate on what our customers expect from apps, to find a way to inspire us with skills from different perspectives and clients. Understanding and discriminating against the psychological and discriminatory people of the participants makes it easier to design apps that will motivate people in different fields. Later, we came up with testing methods that would help break down and identify human variables (Ragunath et al., 2010).

1. **User Input**: Including the clients in the beginning phase of any application or programming configuration assists with revealing a portion of the disadvantages and undesirable highlights that people experience while associating with any applications.
2. **Automated Application Specialist**: Giving a robotised specialist in any application assists with finding the intellectual conduct and the effect of clients' route. This specialist ought to have the option to separate a portion of the obstructions that humans face during the association. Additionally, the specialist should have the option to relate to the people's response and involvement with the application.
3. **Task Investigation**: Breaking down the assignment that is given in the application which clients seem, by all accounts, to be drenched. It very well may be controlled by the occasion's client's taps on specific undertakings.
4. **User Activity**: By dissecting the clients' activity, we can find the client conduct or inclinations. Client activity can help upgrade the application plan. Client activity can be controlled by the quantity of snaps and recurrence of visits by distinguishing the clients' IP addresses (Osama and Aref, 2018).
5. **User Preference**: This strategy can examine customers' premium by enrolling in the application and endorsing it to others. Joining the application can tell that a particular customer has been attracted to the application. Regardless, leaving the application without leaving any information or joining is an awful attitude towards the application as indicated from the customer perspective. Thusly, we can say that the customers are managing issues or are redirected either by finding the application isn't charming, or by finding the application's substance isn't comprehensible and has no strong turn of events (Kim et al., 2003).
6. **Online Review**: The motivation behind the online review is to collect a lot through web to help reveal the human capacities and ability in the

TABLE 4.2

Responses by Majority of the Respondents

Survey Questions of Respondents

Question	Answer
What kind of ambiguity did you experience in past months in technical or in operational way on working with your CRM?	A lot of ambiguities that motivated me in performing a lot better in my job.
What are all confusions you experienced while operating this CRM on which you are operating now?	Lack of direction, inability to recognise different customer needs, whether the customers are satisfied with our services or not.
Was proper documentation available for this CRM?	Yes, it is available for almost all the customers who are suffering issues related to the CRM.
This CRM is based on a prototype model. In this, the prototype is given to clients, and as per requirements of business, they add more new features and deploy final software. As a non-technical or end-user you think this model is fruitful?	Yes, this model is fruitful because it gives a proper idea about the kind of software that is required by the organisation for its smooth functioning and for establishing a better relationship between the customer and the organisation.
What kind of challenges did you face in this kind of model-driven CRM?	Differences in customer needs.
	Lack of initiative from the customer side to establish a positive relationship.
	Corruption and politics in the organisation that prevents the organisation from maintaining a smooth functioning and good relationship with the customers.
	Issues of the customers with each other.
What suggestions will you give as an end-user to software industry that is deploying this kind of CRM?	Make the CRM flexible and dynamic.
	Inform the customers about your new software.
	Give more preference to the emotional aspect of the software compared to the technical part.
	Always try to update your CRM with new and innovative strategies of CRM in order to keep the process interesting and appealing for a longer time.
What suggestions will you give to make this kind of CRM more user-friendly and visually rich?	Make it more people-friendly and less strict and rigid.
Was proper technical training provided from the client side at prior stage?	I have no idea about that.
What challenges did you face when new features are added continuously to CRM?	I was not able to quickly adapt according to the new and innovative features being added in it.
	The new features were not that much appealing for me as I preferred the old and basic ones, which were more understandable and easy to learn.
From a non technical user perspective, what kind of model-driven CRM will you prefer? Please elaborate.	I will prefer a user-friendly CRM, which is less technical and more inclined towards establishing emotional and human relations with people.

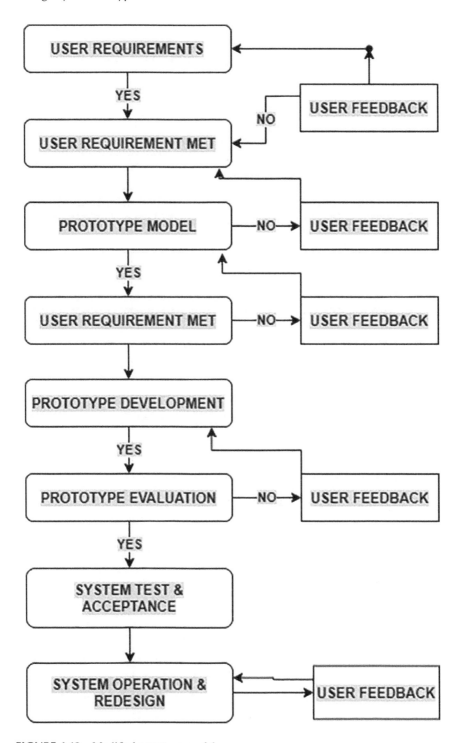

FIGURE 4.12 Modified prototype model.

human-PC connection. It significantly assists with finding the human factors around the world, and locally that furnishes the planner with the outline of the human discernment (Mich and Garigliano, 2000)

4.6.2.2 DANS Software Development Method

One regularly referenced disservice of repetitive working techniques is that they expect groups to begin working right away. Too little thought is given to what in particular precisely is wanted. The assumptions for potential clients or customers are not overseen well. Arrangements concerning the ideal outcomes are insufficient. In this regard, repetitive strategies are less profitable than is the cascade approach, in which these issue are captured comfortably at the start. With an end goal to stay away from this difficulty, DANS applies the better of the two techniques for its product advancement work in Figure 4.13. Tasks start with the cascade technique, so sufficient thought

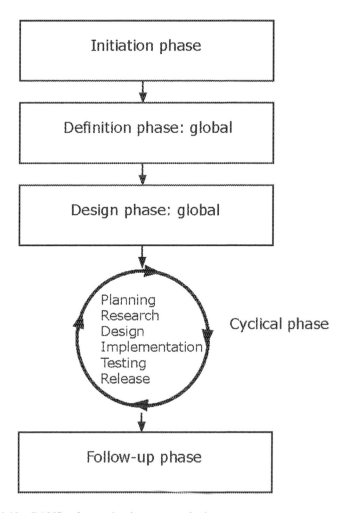

FIGURE 4.13 DANS software development method.

is given to necessities, demands and plan. After the plan stage, there is a shift to the repetitive strategy, subsequently permitting felicity for taking care of these components. The recurrent part of the DANS technique utilises extreme programming (XP). Further definition, plan, execution and testing happen inside the cycles. When the product is adequately evolved, the subsequent stage starts. Each progression in this functioning strategy is depicted underneath (Book_project_management, n.d.).

4.6.2.3   Inception Stage

The inception stage starts with a thought for an undertaking. No financial plan is yet accessible for the undertaking. The objective of this stage is to compose an undertaking plan according to which inside or outer financing can be mentioned.

Exercises in the Inception Stage (Book_project_management, n.d.):
- Elaborate the idea.
- Examine the foundation of help.
- Contact potential accomplices.
- Investigate financing openings.
- Make an underlying worldwide gauge of the manipulated reason for the venture.
- Make a substantial gauge of the reason for the first stage.
- Set project limits.
- Make a venture detail.
- Apply in financing else setting up agreement concurrences with potential clients.

Final Product of the Inception Stage:
- Approved and subsidised undertaking plan.
- Possible understanding with client.

Activities/Decisions:
- Prospective detailed pioneer.
- Client.
- Probable end-user.

4.6.2.4   Definition Stage

Once a partnership has been sponsored, acceptance is the next step. At this stage, the requirements for the consequences of the activity are as clear and predictable as possible. It is about recognising the results of all gatherings. This article or section needs sources or references that appear in credible third-party publications (Book_project_management, n.d.):

- Terms of use.
- Working requirements.
- Operational requirements.
- Design barriers.

4.6.2.5 Configuration Stage

The team can make decisions on the distinctive features of the product by fully defining the requirements in the definition. Program logging is the effect behind the planning phase of IT projects. The program document contains a detailed description of an idea and a general outline of a particular program. The objective product in actual practice and how to explore the same program (e.g. utility level is low). Such a program may be found useful for working with counterfeiters. The forger is not collected immediately; the program is essential for evaluating the program may work only a few or more the construction industry benefits from their completion. In principle, the report supports the plan of measures without hope of any project to modify any selection of the field of fantasy indicates that beyond the first level of producers. Despite the requirements of the archives, these reforms should be able to provide a complete understanding (Book_project_management, n.d.).

Exercises in the plan stage:

- Prepare the plan archive.
- Create and assess models (for example, fakers) with the client.
- Report on the chosen plan.
- Report on the reason that has really been carried out hitherto.
- Make another worldwide gauge of the control factors for the remainder of the task.
- Prepare a substantial gauge of the reason for the iterative stage.

4.6.2.6 Repetitive Stage

The functioning strategies in the repetitive stage are acquired from XP. In this stage, various cycles are acted in progression. A cycle endures from nearly 1 to about 14 days. The accompanying exercises happen inside each cycle (Mazza, 1989; Figure 4.14):

- Deciding.
- Performance test.
- Capacity design.

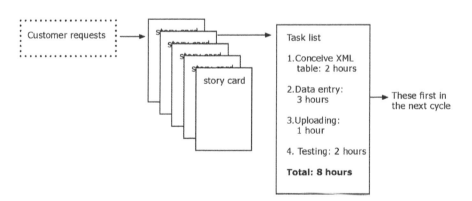

FIGURE 4.14 DANS cyclic story card.

- Implement activities.
- Performance test.
- Sending capacities.

4.7 RESULTS AND DISCUSSION

4.7.1 TOOLS FOR DETECTING AMBIGUITY

Dowser apparatus is an instrument intended to distinguish ambiguities in SRS record utilising parsing method. At first, Dowser parses the necessities utilising compelling syntax. Moreover, an object-situated investigation model of the framework will be created by making classes, techniques, factors and affiliations. Finally, the model will be introduced for the analysts to recognise the equivocalness. Notwithstanding, this procedure doesn't consider recognising vagueness; consequently, the human settles on an ultimate choice of the equivocalness (Matsumoto et al., 2017).

Qualicen is a business apparatus that identifies the conceivable quality deformities such as slice, vague verb modifiers and descriptors, negative words, non-undeniable term, abstract language, imprecise expression, necessities, relative prerequisites, vague pronouns, loopholes, UI detail and long sentence. System distinguishes programming necessities jumble certain prerequisites designing standards utilising POS labelling, morphological investigation and word references. This device displays cautioning messages that contain depiction of the identified smell to the client (Husain and Beg, 2015).

RESI is a tool designed to help programmers. Of course, the archiving time is questionable; the fault provides a framework of exchange that warns of inaccuracies. It provides a potential understanding of every word in the SRS record; therefore, the product expert can change the word. The RESI mechanism identifies functional names that are included in the SRS report and suggests functional terms rather than names. In addition, RESI ensures an adequate dialogue; comparable effects and misrepresentation are widely misused. This is how a RESI instrument works: First, RESI submits the SRS record as a table; second, it checks each word in the SRS report for grammatical meaning (POS). The action word after POS labelling is done consequently, and the framework client can change the labels physically whenever needed. At long last, RESI applies the ontologies WordNet, ResearchCyc, ConceptNet and YAGO to recognise equivocal, flawed and erroneous terms (Alshazly et al., 2014).

SR-Elicitor is a device to computerise the prerequisites elicitation measure, tackle questionable issue in SRS record and produce a controlled portrayal. The specialists of SR-Elicitor utilised Semantics of Business Vocabulary and Business Rules (SBVR) to catch NL SRS report. Figure 4.15 shows the methodology used to make an interpretation of NL programming prerequisites into SBR necessities. Following the interpretation from NL to SBR, SR-Elicitor analyses the NL SRS report. The analysis includes tokenisation, sentence

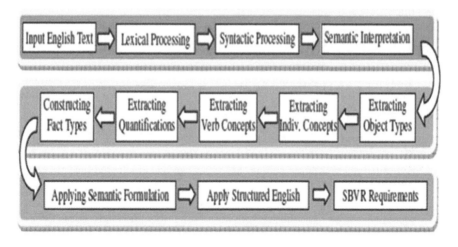

FIGURE 4.15   Process of conversion SRS to tool environment.

breaks, fragmentation and speaking parts (Osman and Zaharin, 2018). The following period of SR-Elicitor instruments is a way to remove SBVR dialect components from data. Since then, the SBVR rules have been created from the SBVR dialects. This step is important to eliminate SBVR requirements and define semantic definitions. The last step for SR-Elicitor is to request a document written in English. In this development, the types of objects are underlined; action word ideas will be in italics; the characters of the SBVR watch will be in bold; and it will divide the ideas of individuals. Figure 4.15 illustrates the requirements of the SBVR (Osama and Aref, 2018).

4.7.1.1   DARA Architecture

This segment gives a compositional portrayal of the DARA framework. It was created to be secluded, extensible and easy to use. We foster a robotised framework to distinguish and resolve ambiguities from full content reports. The DARA engineering is displayed in Figure 4.16. The underlying info is a finished prerequisite book. The yield is unambiguous necessity messages (Sabriye and Wan Zainon, 2018).

4.7.1.2   The Ambiguity-Resolving Module

At long last, this module centres in eliminating and settling the vagueness. For each vague sentence, resolve the uncertainty in the sentence consequently as the last advance utilising settling rules, and along these lines, further develop the normal language prerequisite specification document. Figure 4.4 shows the ambiguity-resolving module engineering (Osama and Aref, 2018).

The settling uncertainty approach utilises the accompanying normal guidelines to check if a sentence contains vagueness (Sabriye and Wan Zainon, 2018).

Rule 1: In any case, between the two at the same time, outside in addition, while participating, however I repeat it in two sentences (Sabriye and Zainon, 2017).

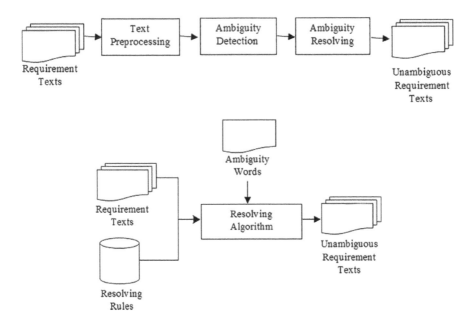

FIGURE 4.16 Detection module architecture.

Rule 2: At the point when sentence containing aside from if, override with if not (Matsumoto, et al., 2017).

Rule 3: At the point when sentence containing a, an, all, any, a couple, every, couple of replace with each.

Rule 4: When sentence containing ought to, will, would, may, might, should supplant with will.

Rule 5: At the point when sentence containing there is X in Y, X exists in Y override with Y has X.

Rule 6: At the point when sentence containing anaphora or pronoun, for instance, they or them replaces with the farthest thing.

Rule 7: When phrase assume that connection with each one of which.

Rule 8: When sentence containing just, additionally, nearly, even, barely, just, simply, almost, and truly put first action word (Alshazly et al., 2014).

Rule 9: When sentence containing until, up to, at, during, span and including, through, by, or after add just before it (Haron and Ghani, 2015).

Rule 10: Phrase comprise and, or in same sentence add parentheses.

Rule 11: Phase comprises many supplant with every one of many.

Rule 12: Phrase comprise not many supplant with every one of few.

Rule 13: Phrase carries for up to supplant with for up to and including.

Rule 14: Phrase carries plural things add each before it (Osman and Zaharin, 2018).

Figure 4.17 depicts that some prominent ambiguities are more often recognised than others by particularly lexical, extension and obscure vagueness is by means

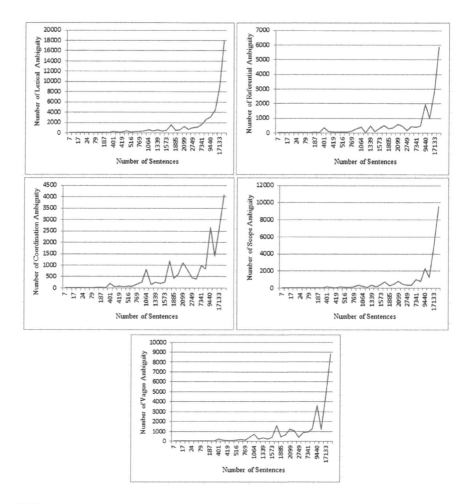

FIGURE 4.17   No. of frequencies detected in ambiguity.

the ambiguity is predicted. Figure 4.6 shows that record 3 exhibits a decline in rate dissemination of all equivocalness types distinguished in light of the report domain (Document 3 about satellite) and it shows that archive 26 shows an expansion in rate conveyance of all vagueness types identified due to the report area canvassed in word references (Haron and Ghani, 2015).

4.7.2   Risk Analysis Due to Ambiguity in Requirements

The likelihood of a risk happening can be assessed dependent on a few components as dictated by the extraordinary idea of each task. For instance, factors assessed for potential H/W or S/W innovation dangers could incorporate the innovation not being adult, the innovation being excessively unpredictable and a deficient help base for fostering the innovation (Nigam et al., 2012). The effect of a danger happening could

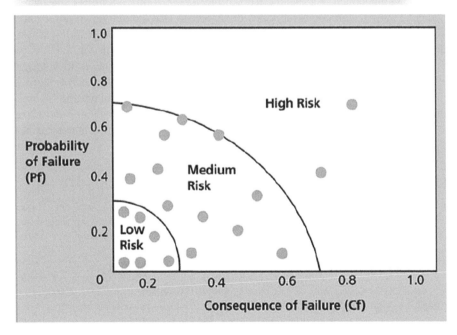

FIGURE 4.18   Probability distribution of risk.

incorporate factors such as accessibility of fallback arrangements or the outcomes of not gathering execution, cost and timetable evaluations (Husain and Beg, 2015).

The above figure 4.18 gives an illustration of how the danger factors were utilised to chart the likelihood of disappointment and result of disappointment for the proposed advances. The figure groups potential innovations (dabs on the outlines) as high, medium, or generally safe dependent on the likelihood of disappointment and outcome of disappointment. The analysts for this investigation strongly suggested that the US Air Force put resources into the low- to medium-danger advances and proposed that it does not seek after the high-danger innovations. It tends to be seen that the meticulousness behind utilising probability/impact matrix and hazard factors gives a lot more grounded contention than just expressing the danger probabilities, or results are high, medium or low (Sabriye and Zainon, 2017).

4.8 CONCLUSIONS

This chapter makes two commitments to lessen the degree of uncertainty in modern necessities reports. To start with, it offers to a prerequisites engineer a productive assessment procedure for distinguishing questionable necessities that are material in mechanical RE. Second, it offers a way to deal with distinguishing equivocalness types that can happen in a specific RE setting (Baars, 2006). All things considered, one can't anticipate recognising kinds of vagueness that the individual never under any circumstance has pondered or run over (Sabriye and Wan Zainon, 2018). Maybe, the commitment lies in the methodical method to investigate this verifiably existing information by utilising the heuristics and in expanding the necessities designer's attention to the issue. Our future work targets examining how much gatherings increase the quantity of recognised ambiguities. In gatherings, maybe ambiguities that have fallen through singular readiness can be identified (Abduljalil and Kang, 2011). We should figure out which meeting designs permit commentators to best trade their understandings of prerequisites (Gupta et al., 2019).

REFERENCES

Abduljalil, S., & Kang, D.-K. (2011). Effective Model and Methods for Analysing Human Factors in Software Design for Efficient User Experience. *Journal of Information and Communication Convergence Engineering*, 9(1), 100–104. https://doi.org/10.6109/jicce.2011.9.1.100.

Alferoff, C., & Knights, D. (2008). Customer relationship management in call centers: The uneasy process of re(form)ing the subject through the "people-by-numbers" approach. *Information and Organization*, 18(1), 29–50. https://doi.org/10.1016/j.infoandorg.2007.10.002.

Alokla, M., Alkhateeb, M., Abbad, M., & Jaber, F. (2019). Customer relationship management: A review and classification. *Transnational Marketing Journal*, 7(2), 187–210. https://doi.org/10.33182/tmj.v7i2.734.

Alshazly, A. A., Elfatatry, A. M., & Abougabal, M. S. (2014). Detecting defects in software requirements specification. *Alexandria Engineering Journal*, 53(3), 513–527. https://doi.org/10.1016/j.aej.2014.06.001.

Baars, W. (2006). Project management handbook DANS (Data Archiving and Networked Services). *Proceedings – 2014 Agile Conference, AGILE 2014, July*, 84. http://localhost:8080/ccsdsdocs/LIFECYCLE-ProjectManagement-DANS-en.pdf.

Bano, M. (2016). Addressing the challenges of requirements ambiguity: A review of empirical literature. *5th International Workshop on Empirical Requirements Engineering, EmpiRE 2015 – Proceedings, July*, 21–24. https://doi.org/10.1109/EmpiRE.2015.7431303.

Bäumer, F. S., & Geierhos, M. (2018). Flexible ambiguity resolution and incompleteness detection in requirements descriptions via an indicator-based configuration of text analysis pipelines. *Proceedings of the 51st Hawaii International Conference on System Sciences, 9*, 5746–5755. https://doi.org/10.24251/hicss.2018.720.

Chen, I. J., & Popovich, K. (2003). Understanding customer relationship management (CRM): People, process and technology. *Business Process Management Journal, 9*(5), 672–688. https://doi.org/10.1108/14637150310496758.

Dalpiaz, F., van der Schalk, I., Brinkkemper, S., Aydemir, F. B., & Lucassen, G. (2019). Detecting terminological ambiguity in user stories: Tool and experimentation. *Information and Software Technology, 110*(December), 3–16. https://doi.org/10.1016/j.infsof.2018.12.007.

Ezzini, S., Abualhaija, S., Arora, C., Sabetzadeh, M., & Briand, L. C. (2021). Using domain-specific corpora for improved handling of ambiguity in requirements. *2021 IEEE/ACM 43rd International Conference on Software Engineering (ICSE)*. 1485–1497. https://doi.org/10.1109/icse43902.2021.00133.

Fantechi, A., Ferrari, A., Gnesi, S., & Semini, L. (2018). Hacking an ambiguity detection tool to extract variation points: An experience report. *ACM International Conference Proceeding Series*, 43–50. https://doi.org/10.1145/3168365.3168381.

Ferrari, A., Lipari, G., Gnesi, S., & Spagnolo, G. O. (2014). Pragmatic ambiguity detection in natural language requirements. *2014 IEEE 1st International Workshop on Artificial Intelligence for Requirements Engineering, AIRE 2014 – Proceedings, August*, 1–8. https://doi.org/10.1109/AIRE.2014.6894849.

Ganpatrao Sabale, R. (2012). Comparative study of prototype model for software engineering with system development life cycle. *IOSR Journal of Engineering, 02*(07), 21–24. https://doi.org/10.9790/3021-02722124.

Gervasi, V., Ferrari, A., Zowghi, D., & Spoletini, P. (2019). Ambiguity in requirements engineering: Towards a unifying framework. *Lecture Notes in Computer Science (Including Subseries Lecture Notes in Artificial Intelligence and Lecture Notes in Bioinformatics), 11865*, 191–210. https://doi.org/10.1007/978-3-030-30985-5_12.

Grieskamp, W., Heisel, M., & Dörr, H. (1998). Specifying embedded systems with statecharts and Z: An agenda for cyclic software components. *Lecture Notes in Computer Science (Including Subseries Lecture Notes in Artificial Intelligence and Lecture Notes in Bioinformatics), 1382*, 88–106. https://doi.org/10.1007/bfb0053585.

Gupta, A. K., Deraman, A., & Siddiqui, S. T. (2019). A survey of software requirements specification ambiguity. *ARPN Journal of Engineering and Applied Sciences, 14*(17), 3046–3061.

Hammer, J. H., & Vogel, D. L. (2013). Assessing the utility of the willingness/prototype model in predicting help-seeking decisions. *Journal of Counseling Psychology, 60*(1), 83–97. https://doi.org/10.1037/a0030449.

Haron, H., & Ghani, A. A. A. (2015). A survey on ambiguity awareness towards malay system requirement specification (SRS) among industrial IT practitioners. *Procedia Computer Science, 72*, 261–268. https://doi.org/10.1016/j.procs.2015.12.139.

Hayman Oo, K., Nordin, A., Ritahani Ismail, A., & Sulaiman, S. (2018). An analysis of ambiguity detection techniques for software requirements specification (SRS). *International Journal of Engineering & Technology, 7*(2.29), 501. https://doi.org/10.14419/ijet.v7i2.29.13808.

Husain, M. S., & Beg, M. R. (2015). Advances in ambiguity less NL SRS : A review. *2015 IEEE International Conference on Engineering and Technology (ICETECH) March 2015*, 221–225.

Kamsties, E., Berry, D. M., & Paech, B. (2001). Detecting ambiguities in requirements documents using inspections. *Engineering, July*, 1–12. http://publica.fraunhofer.de/documents/

N-6838.html%5Cnhttp://www.cas.mcmaster.ca/wise/wise01/KamstiesBerryPaech. pdf%5Cnhttp://citeseerx.ist.psu.edu/viewdoc/download?doi=10.1.1.145.6497&re p=rep1&type=pdf.

Kim, J., Suh, E., & Hwang, H. (2003). A model for evaluating the effectiveness of CRM using the balanced scorecard. *Journal of Interactive Marketing*, *17*(2), 5–19. https://doi. org/10.1002/dir.10051.

Massey, A. K., Rutledge, R. L., Antón, A. I., & Swire, P. P. (2014). Identifying and classifying ambiguity for regulatory requirements. *2014 IEEE 22nd International Requirements Engineering Conference, RE 2014 – Proceedings*, *115*(111), 83–92. https://doi. org/10.1109/RE.2014.6912250.

Matsumoto, Y., Shirai, S., & Ohnishi, A. (2017). A method for verifying non-functional requirements. *Procedia Computer Science*, *112*, 157–166. https://doi.org/10.1016/j. procs.2017.08.006.

Mazza, C. (1989). Software project management. *Computer Physics Communications*, *57*(1–3), 23–28. https://doi.org/10.1016/0010-4655(89)90188-4.

Mich, L., & Garigliano, R. (2000). Ambiguity measures in requirements engineering. *ICS2000 International Conference on Software: Theory and Practice. 16th IFIP World Computer Congress*, 39–48.

Nacheva, R. (2017). Prototyping approach in user interface. *2nd Conference on Innovative Teaching Methods, June*, 80–87. https://www.researchgate.net/publication/317414969.

Nigam, A., Arya, N., Nigam, B., & Jain, D. (2012). Tool for automatic discovery of ambiguity in requirements. *International Journal of Computer Science Issues (IJCSI)*, *9*(5), 350–356. www.IJCSI.org.

Osama, S., & Aref, M. (2018). Detecting and Resolving Ambiguity Approach in Requirement Specification: Implementation, Results and Evaluation. *International Journal of Intelligent Computing and Information Sciences*, *18*(1), 27–36. https://doi. org/10.21608/ijicis.2018.15909

Osman, M. H., & Zaharin, M. F. (2018). Ambiguous software requirement specification detection: An automated approach. *Proceedings - International Conference on Software Engineering*, 33–40. https://doi.org/10.1145/3195538.3195545

Pittke, F., Leopold, H., & Mendling, J. (2016). Automatic detection and resolution of lexical ambiguity in process models (extended abstract). *Lecture Notes in Informatics (LNI), Proceedings - Series of the Gesellschaft Fur Informatik (GI)*, *P252*, 75–76.

Ragunath, P., Velmourougan, S., Davachelvan, P., Kayalvizhi, S., & Ravimohan, R. (2010). Evolving a new model (SDLC Model-2010) for software development life cycle (SDLC). *International Journal of Computer Science and Network Security*, *10*(1), 112–119.

Rauterberg, M., Strohm, O., & Kirsch, C. (1995). Benefits of user-oriented software development based on an iterative cyclic process model for simultaneous engineering. *International Journal of Industrial Ergonomics*, *16*(4–6), 391–409. https://doi. org/10.1016/0169-8141(95)00021-8.

Sabriye, A. O. J., & Wan Zainon, W. M. N. (2018). An approach for detecting syntax and syntactic ambiguity in software requirement specification. *Journal of Theoretical and Applied Information Technology*, *96*(8), 2275–2284.

Sabriye, A. O. J, & Zainon, W. M. N. W. (2017). A framework for detecting ambiguity in software requirement specification. *ICIT 2017–8th International Conference on Information Technology, Proceedings, May 2017*, 209–213. https://doi.org/10.1109/ICITECH.2017. 8080002.

Yang, H., Willis, A., De Roeck, A., & Nuseibeh, B. (2010). Automatic detection of nocuous coordination ambiguities in natural language requirements. *ASE '10: Proceedings of the IEEE/ACM international conference on Automated Software Engineering*. 53. https:// doi.org/10.1145/1858996.1859007.

5 Selection of Software Programmer Using Fuzzy MCDM Technique in Software Engineering Scenario

Ragini Shukla
Dr. C. V. Raman University

CONTENTS

5.1 INTRODUCTION

In the software engineering area, effort estimation is undefined and it's based on different external elements while producing a certain type of software. Software development companies must choose an ideal and experienced group of developers for organisational benefits. Because the success or failure of software is primarily dependent on knowledgeable persons, this is required.

To get the best effort from programmers in any software concern, this is the objective and is obtained by this study where the ranking of programmers among groups of programmers is carried out. The appraisal of programmer's ranking is one of

DOI: 10.1201/9780367816414-5

the most crucial tasks, having complex and conjugate outcomes of human nature such as experience, knowledge and skills. To get programmers' rank, I tried to adopt some features of the programmers, viz. basic skills, communication between colleagues, logic analysis capability and, most crucially, how much experience they hold. These are the parameters that I get from experts and renowned personalities who hold extensive evaluation knowledge in relevant fields. In this chapter, efforts have been made to reach such conclusions where this gives the best decision which is acceptable with simple and most acceptable methodology; further, more discussions are also there, where it covers this topic in an elaborate manner.

This chapter describes the integration of MCDM-based FAHP and FTOPSIS methods that are used in the creation or selection of software development team. Section 5.2 shows the review of ranking-based optimisation techniques, and Section 5.3 describes the criteria and alternatives of the programmer. Section 5.4 shows fuzzy MCDM techniques, Section 5.5 the evaluation of programmers' rank using FAHP, Section 5.6 the appraisal of programmers' rank using integrated FAHP and FTOPSIS, and Section 5.7 comparative analysis. This chapter concludes in Section 5.8 with some details concerning the evaluation of these two methods.

5.2   REVIEW OF RANKING-BASED OPTIMISATION TECHNIQUES

A number of researchers have worked in this field. A short resume of activities and developments in the field is been given below.

Some researchers have focused on MCDM-based AHP techniques for ranking-based estimation. In this context, Pogarcic [1] looked into the possibility of using AHP to make decisions on traffic planning and implementation, as well as ensuring high-quality business logistics. Mishra [3] also built a selection algorithm based on expert evaluations that combines AHP and Bayesian networks to choose the most efficient developers. It also determines the best order for developers based on their capabilities, as well as the number of developers to choose from based on sensitivity values.

There are many situations when numeric data will not be available and the data are fuzzy in nature. In this case, we can use the FAHP as ranking-based estimation. Many authors have used FAHP technique instead of AHP technique to incorporate the fuzzy nature of variables.

Yuen [4] suggested a fuzzy AHP model for estimating software quality and choosing software vendors in the face of uncertainty. The model employs a fuzzy logarithmic least squares method that has been updated. This model's usability and validity are defined by an arithmetic example. Gungor et al. [5] identified the best appropriate person and established the MCDM model. They also proposed Yager's weighted technique, which they compared to the FAHP method's results. Finally, based on these findings, the FAHP technique and Yager's weighted method both recommend the same option as the optimal option.

Buyukozkan et al. [7] gave an approach for evaluating the performance of operators. Buyukozkan [8] established a model for evaluating service quality in the healthcare sector, as well as the performance of select pioneer Turkish hospitals, using the quality factors. The FAHP was used, and the results revealed that hospitals should place a greater emphasis on empathy, professionalism and dependability in order to provide satisfying and quality service.

Catak [9] created a novel fuzzy AHP-based decision model that may be used to quickly select a database management system. This study demonstrates that choosing a database management system does not have to be difficult and is one of the most critical operations in a company's IT project.

Javanbarg et al. [10] presented a basic fuzzy optimisation model for FAHP-based MCDM system. They suggested fuzzy prioritising approach can generate crisp priorities from both consistent and inconsistent pair-wise assessments using nonlinear optimisation model. The judgments are represented as triangular fuzzy numbers in the proposed nonlinear optimisation approach, which eliminates the need for an additional aggregation procedure.

FAHP is an extension of AHP; a comparative study is also needed to check the performance of one over the other. In this context, Kabir [12] offered a relative analysis of AHP and FAHP for multi-criteria inventory classification model. Sehra et al. [13] examined the application of the FAHP method of MCDM for selecting the best model based on the company atmosphere and type of the project, and it gave better results compared to AHP.

Another MCDM technique that is popularly used for ranking is TOPSIS. Very few studies are available on this technique. Bondor and Muresan [14] discussed the problem of decision-making. The proposed method can tackle the problem of multicollinearity between criteria. The goal of their technique was to use the TOPSIS method repeatedly until the correlations between components were minimised to a certain level.

Wimatsari et al. [16] using the fuzzy MCDM technique for TOPSIS were able to achieve scholarship recipient selection results. Based on value choices, the selection recommends an alternative with the highest level of eligibility to the least eligible for a scholarship.

Fuzzy AHP and fuzzy TOPSIS are the most popular MCDM methods. The technique of blending these two techniques is very crucial for ranking-based estimation. The proposed optimisation process needs no exhaustive computation; it doesn't matter whether it is computation of m-dimensional eigenvector or it is calculation of m by n manipulated fuzzy ratings. This process of fuzzy ranking type of decision-making justifies with a note that the methodology works fine for the variety of criteria expressed either in mathematically crisp form or in linguistic form. In the proposed methodology, the computation cost is minimum and there is no use of weight, which is a lengthy process to find out.

5.3 EFFORT MULTIPLIERS AS CRITERIA AND ALTERNATIVE IN SOFTWARE ENGINEERING SCENARIO

In an instant work, selected alternatives and criteria of programmer are the two factors importantly applied in the MCDM method. First, we select alternative and criteria because both the factors are initially very important for this method. In this method, AHP and TOPSIS techniques are the greatest suited methods for obtaining and selecting the best alternatives; for instance, programmers are alternatives and skill, knowledge, experience, etc., are criteria.

In this research work, the selection of criteria plays the crucial role; it is the criterion on which end results depend. An efficient and experienced group of software programmers may highly influence the accuracy of effort estimation; for this, a suitable and reliable technique is required, which will select the best group of programmers based on some criteria. The constructive cost model (COCOMO) [13] is one of the popular and reliable methods of software effort estimation created on 17 effort multipliers; these are fuzzy in nature. This research work utilises these 17 effort multipliers as criteria to be applied with FAHP and FTOPSIS methods to select the finest group of programmers. These multipliers as shown in Table 5.1 are rated on a scale – very low to extra high. We can observe that all multipliers are not only quantitative, but also qualitative quantifiers. We cannot assign a precise value for them; hence, fuzzy-based ranking method is needed to assign and evaluate imprecise value for each multiplier.

At first, three multipliers out of 17 [14]: APEX – application experience, PLEX – platform experience and LTEX – language and tool experience, are considered as a criteria for the FAHP method [18]. As discussed, FAHP techniques are applied in three initial multipliers; it is also possible to apply in a larger no. of criteria and alternatives. After that, I considered all 17 effort multipliers as criteria for the FAHP and FTOPSIS methods. A sample data of ten software programmers are considered in fuzzy terms and applied on two MCDM methods. Ranks obtained through FAHP and FTOPSIS are compared and found to be satisfactory.

TABLE 5.1
COCOMO'81 Dataset Statistics

S. No.	Variable	Type	Description
1	Acap	Numeric	Analysts capability
2	Pcap	Numeric	Programmers capability
3	Apex	Numeric	Application experiences
4	Modp	Numeric	Modern programming practices
5	tool	Numeric	Use of software tools
6	vexp	Numeric	Virtual machine experience
7	lexp	Numeric	Language experience
8	sced	Numeric	Schedule constraint
9	stor	Numeric	Main memory constraint
10	data	Numeric	Database size
11	time	Numeric	Time constraint for CPU
12	turn	Numeric	Turnaround time
13	virt	Numeric	Machine volatility
14	cplx	Numeric	Process complexity
15	rely	Numeric	Required software reliability
16	loc	Numeric	Line of code
17	effort	Numeric	Overall effect

5.4   FUZZY MCDM

The theory of decision-making established a foundation for better ordered and reasonable decision-making, particularly in situations when numerous factors must be considered MCDM. In many ways, the decision-making problems are identical and ambiguous. Zadeh's introduction to the uncertainty theory of fuzzy set has been received by MCDM enthusiasts. The merging of MCDM and fuzzy set theory strengthens the fuzzy MCDM, a new decision theory. The fuzzy MCDM approaches have been used in a variety of real-world situations.

5.4.1   FAHP

The FAHP method [19], which is derived from the AHP, is a more advanced analytical method. Despite its widespread use, the AHP has been chastised for failing to address the inherent ambiguity and imprecision that come with mapping a decision-perspective makers to precise numbers. In order to allow ambiguity, the FAHP technique uses fuzzy comparisons ratios [9]. Chen and Hwang (1992) devised a system that turns linguistic terms into fuzzy numbers before converting them back to linguistic terms.

   i. **Demonstration of the Method**: Now, the five-point scale is considered to demonstrate the conversion of fuzzy number into crisp scores. To demonstrate the method, a five-point scale having the linguistic terms such as low, below average, average, above average and high as shown in Figure 5.1 is considered.

The main procedure of AHP is as follows [14].

   Step 1: Determine the attributes and objective.
   Step 2: Assess the relative significance of various features in relation to the aim or objective and the rating based on Saaty's nine-point scale.

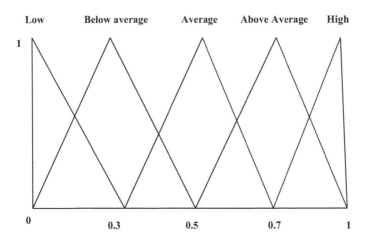

FIGURE 5.1   Conversion of fuzzy numbers to fuzzy terms fuzzification of linguistic terms.

- Find the relative normalised weight (w_j) of each attribute by
  ii) Normalizing the geo metric means of rows in the comparison matrix.

$$GM_j = \left[\prod_{i=1}^{M} b_{ij} \right]^{1/M} \text{ and } w_j = GM_j \Big/ \sum_{i=1}^{M} GM_j \qquad (5.1)$$

- Calculate matrices $A3$ and $A4$ such that

$$A3 = A1 \times A2 \qquad (5.2)$$

$$\text{and } A4 = A3 / A2, \qquad (5.3)$$

where $A2 = [w_1, w_2 \ldots, w_i]^T$.
- Determine the maximum eigenvalue $\lambda_{\max}$ that is the average of matrix $A4$.
- Calculate the consistency index $CI = (\lambda_{\max} - M) \Big/ (M - 1)$.
- Obtain the random index (RI) for the number of attributes used in decision-making. Calculate the consistency ratio $CR = CI/RI$. Usually, a CR of 0.1 or less is considered acceptable and it reflects an informed judgement attributable to the knowledge of the analyst regarding the problem under study.

Step 3: The next stage is to pair-wise compare the alternatives to see how much better they are at meeting each of the criteria.

Step 4: The last step is to acquire the overall scores for the alternatives.

5.4.2 FUZZY TECHNIQUE FOR ORDER PREFERENCE BY SIMILARITY TO IDEAL SOLUTION (FTOPSIS)

TOPSIS can identify solutions from a finite set of alternatives [22]. The logic of fuzzy TOPSIS according to Hwang and Yoon (1981) is to define the positive ideal solution and negative ideal solution. The positive ideal solution is the solution that maximises the benefit metrics and minimises the cost metrics, whereas the negative ideal solution is the solution that maximises the cost metrics and minimises the benefit metrics. The best alternative is the one which has the shortest distance from the positive ideal solution and the farthest distance from the negative ideal solution. But it is often difficult for a decision-maker to assign a precise performance rating to an alternative for the attributes under consideration. Then the merit of using a fuzzy TOPSIS approach is to assign different metric values using fuzzy numbers. In this study, the AHP is used to analyse the structure selection problem and to determine weights of the criteria, and fuzzy TOPSIS method is used to obtain final ranking.

The general implementing steps from fuzzy TOPSIS procedure for multi-criteria group decision-making to make some modified calculations are as follows:

Step 1: More importantly, first, the Board (including the members who make decisions) is formed and the evaluation criteria are identified.

Step 2: Declare the appropriate variable in specific language for different criteria to make weight important and also provide rating for alternative in respect of criteria.

Step 3: To get the cumulative fuzzy or as an aggregate weight, aggregate the weight criteria.

Step 4: Construct the fuzzy matrix in a normalised form.

$$r_{ij} = \frac{x_{ij}}{\left[\sum_i x_{ij}^2\right]^{\frac{1}{2}}},$$ (5.4)

for $i = 1,\ldots, m; j = 1,\ldots, n$.

Step 5: Construct the weighted fuzzy decision matrix v_{ij} in a normalised form. This is done by the multiplication of each element of the column of the matrix R_{ij} with its associated weight W_j. Hence, normalised matrix v_{ij} is expressed as:

$$v_{ij} = w_j \,^{\circ} r_{ij}$$ (5.5)

Step 6: Obtain the ideal (best) and negative ideal (worst) solutions in this step:
Ideal solution:
$A^* = \{v_1^*, \ldots, v_n^*\}$, where

$$v_j^* = \left\{ \max_i \left(v_{ij}\right) \text{if } j \in J; \min_i \left(v_{ij}\right) \text{if } j \in J'\right\}$$ (5.6)

Negative ideal solution:
$A' = \{v_1', \ldots, v_n'\}$, where

$$v_j' = \left\{ \min_i \left(v_{ij}\right) \text{if } j \in J; \max_i \left(v_{ij}\right) \text{if } j \in J'\right\}$$ (5.7)

Step 7: It is needed in this point to calculate the fuzzy parameters.

Step 8: Calculate each alternative's coefficients and its closeness.
The separation from the ideal alternative is:

$$S_i^* = \left[\sum_j \left(v_j^* - v_{ij}\right)^2\right]^{\frac{1}{2}}; i = 1,\ldots, m$$ (5.8)

Similarly, the separation from the negative ideal alternative is:

$$S_i' = \left[\sum_j \left(v_j' - v_{ij}\right)^2\right]^{\frac{1}{2}}; i = 1,\ldots, m$$ (5.9)

Step 9: As per closeness co-efficiency, order the rank of all alternatives.
The relative closeness to the ideal solution C_i^* is

$$C_i^* = \frac{S_i'}{S_i^* + S_i'}$$ (5.10)

These are the overall steps of TOPSIS through which we can decide the rank.

5.4.3   INTEGRATED **FAHP** AND **FTOPSIS** METHOD

We use two-step methods consisting of FAHP and FTOPSIS; in the first step, the FAHP is used for calculating the weights of the attributes or criteria as well as the overall weights of the candidates in each attribute. In the second step, these weights are considered and used in the FTOPSIS process. Then FTOPSIS is applied for the evaluation problem, and the result shows the preference order of the programmer. These methodology levels can be discussed clearly, and their steps are shown in below.

FAHP:

Step 1: Determine the objective.
Step 2: Select experts and attributes/criteria and identify the alternatives.
Step 3: Establish the pair-wise comparison matrix of the criteria.
Step 4: Derive the eigenvalue and eigenvector.
Step 5: Perform the consistency test.
Step 6: Compute the weights of the criteria.
Step 7: Establish the pair-wise comparison of the alternatives with respect to each criterion.
Step 8: Perform the consistency test.
Step 9: Compute the weights of the alternatives for each criterion.
Step 10: Calculate the geometric mean of the weights calculated by experts.
Step 11: Calculate the eigenvalue and eigenvector.
Step 12: Perform the consistency test.
Step 13: Compute the overall weights of the alternative.

FTOPSIS:

Step 14: Start TOPSIS procedure using the weights calculated using the AHP.
Step 15: Calculate negative and positive ideal solutions and separation measures.
Step 16: Rank the preference candidate in descending order.

5.5   EVALUATION OF PROGRAMMERS' RANK USING FAHP

The traditional AHP method is problematic, because this method shows exact values to express the decision-makers' opinion in a comparison of alternatives. In spite of the traditional AHP method, the study of fuzzy AHP is used to compare fuzzy ratios described by triangular fuzzy numbers. Chang (1991) introduced a new approach for handling fuzzy AHP, here the use of triangular fuzzy numbers for pairing comparison scale of fuzzy AHP in pair-wise system. Figure 5.2 shows the hierarchy of programmer selection (Prog 1, Prog 2, Prog 3) [14]. In order to apply the FAHP method, the steps below are followed:

Step 1: Constructed a DMM based on the above attribute with three fuzzy linguistic terms.

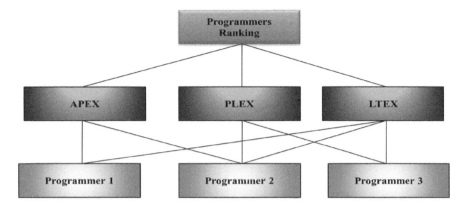

FIGURE 5.2   Hierarchical threshold levels.

From the Chen and Hwang (1992) method, the fuzzy linguistic term is converted into crisp data using three-point scale.

Step 2: Now in this step, we compare criteria with criteria by assigning comparative weights

$$
\begin{array}{c}
\text{APEX} \\
\text{PLEX} \\
\text{LTEX}
\end{array}
\begin{bmatrix}
\text{APEX} & \text{PLEX} & \text{LTEX} \\
1 & 5 & 3 \\
1/5 & 1 & 1/2 \\
1/3 & 2 & 1
\end{bmatrix}
$$

Now calculating geometric mean (GM) for ith row: $GM_1 = (1 \times 5 \times 3)^{1/3} = 2.4659$, $GM_2 = (1/5 \times 1 \times 1/2)^{1/3} = 0.4641$, and $GM_3 = (1/3 \times 2 \times 1)^{1/3} = 0.873$.

The total geometric mean GM = 3.79.

Hence, the normalised weights are: $W_1 = 2.46/3.79 = 0.649$, $W_2 = 0.46/3.79 = 0.121$, and $W_3 = 0.87/3.79 = 0.229$.

Now consistency checking

$$
A_3 = \begin{bmatrix} 1 & 5 & 3 \\ 1/5 & 1 & 1/2 \\ 1/3 & 2 & 1 \end{bmatrix} \times \begin{bmatrix} 0.649 \\ 0.121 \\ 0.229 \end{bmatrix} = \begin{bmatrix} 1.914 \\ 0.36 \\ 0.678 \end{bmatrix}
$$

$$
A_4 = \begin{bmatrix} 1.914 \\ 0.36 \\ 0.678 \end{bmatrix} \div \begin{bmatrix} 0.649 \\ 0.121 \\ 0.229 \end{bmatrix} = \begin{bmatrix} 2.949 \\ 2.975 \\ 3.081 \end{bmatrix}
$$

And the maximum value $\lambda_{\max}$ that is the average of matrix A_4 will be

$$
\lambda_{\max} = \frac{2.949 + 2.975 + 3.0818}{3} = 3.001
$$

Then the consistency index (CI) $= \dfrac{(\lambda_{max} - n)}{n-1} = \dfrac{3.001 - 3}{2} = 0.0005.$

And the consistency ratio (CR) $= \dfrac{CI}{RI} = \dfrac{0.0005}{0.52} = 0.00096 < 0.1.$

Hence, the weights are consistent.

Step 3: Now alternatives will be compared with alternatives known as pairwise comparison matrix.

i. For criteria APEX

	Prog1	Prog2	Prog3
Prog1	1	0.495	0.895
Prog2	1 / 0.495	1	0.895
Prog3	1 / 0.895	1 / 0.895	1

Now calculating geometric mean (GM) for ith row: $GM_1 = (1 \times 0.495 \times 0.895)^{1/3} = 0.7623$, $GM_2 = (1/0.495 \times 1 \times 0.895)^{1/3} = 1.2182$, and $GM_3 = (1/0.895 \times 1/0.895 \times 1)^{1/3} = 1.0767$.

By equations (5.2) and (5.3) as below:

$$A_3 = \begin{bmatrix} 1 & 0.495 & 0.895 \\ 1/0.495 & 1 & 0.895 \\ 1/0.895 & 1/0.895 & 1 \end{bmatrix} \times \begin{bmatrix} 0.249 \\ 0.398 \\ 0.352 \end{bmatrix} = \begin{bmatrix} 0.7614 \\ 1.2167 \\ 1.0752 \end{bmatrix}$$

$$\text{And } A_4 = \begin{bmatrix} 0.7614 \\ 1.2167 \\ 1.0752 \end{bmatrix} \div \begin{bmatrix} 0.249 \\ 0.398 \\ 0.352 \end{bmatrix} = \begin{bmatrix} 3.074 \\ 3.005 \\ 3.053 \end{bmatrix}$$

And the maximum value λ_{max} that is the average of matrix A_4:

$$\lambda_{max} = \frac{3.074 + 3.005 + 3.053}{3} = 3.044$$

Then CI $= \dfrac{(\lambda_{max} - n)}{n-1} = \dfrac{3.044 - 3}{2} = 0.022.$

And CR $= \dfrac{CI}{RI} = \dfrac{0.022}{0.52} = 0.04 < 0.1.$

Hence, the weights are consistent.

ii. For criteria PLEX

	Prog1	Prog2	Prog3
Prog1	1	0.895	0.115
Prog2	1 / 0.895	1	0.115
Prog3	1 / 0.115	1 / 0.115	1

Now calculating geometric mean (GM) for ith row: $GM_1 = (1 \times 0.895 \times 0.115)1/3 = 0.4686$, $GM_2 = (1/0.895 \times 1 \times 0.115)1/3 = 0.50464$, and $GM_3 = (1/0.115 \times 1/0.115 \times 1)1/3 = 4.2280$.

Total GM $= 5.2012$.

Hence, the normalised weights are: $W_1 = 0.4686/5.2012 = 0.090$, $W_2 = 0.50464/5.2012 = 0.0970$, and $W_3 = 4.2280/5.2012 = 0.81288$.

Now consistency checking

$$A_3 = \begin{bmatrix} 1 & 0.895 & 0.115 \\ 1/0.895 & 1 & 0.115 \\ 1/0.115 & 1/0.115 & 1 \end{bmatrix} \times \begin{bmatrix} 0.090 \\ 0.0970 \\ 0.812 \end{bmatrix} = \begin{bmatrix} 0.2701 \\ 0.2908 \\ 2.438 \end{bmatrix}$$

$$\text{And } A_4 = \begin{bmatrix} 0.2701 \\ 0.2908 \\ 2.438 \end{bmatrix} \div \begin{bmatrix} 0.090 \\ 0.0970 \\ 0.812 \end{bmatrix} = \begin{bmatrix} 3.001 \\ 2.997 \\ 3.002 \end{bmatrix}$$

And the maximum value $\lambda_{\max}$ that is the average of matrix A_4:

$$\lambda_{\max} = \frac{3.001 + 2.997 + 3.002}{3} = 3$$

Then CI $= \dfrac{(\lambda_{\max} - n)}{n-1} = \dfrac{3-3}{2} = 0$.

And CR $= \dfrac{CI}{RI} = \dfrac{0}{0.52} = 0 < 0.1$.

Hence, the weights are consistent.

iii. For criteria LTEX

$$\begin{array}{c} \text{Prog1} \\ \text{Prog2} \\ \text{Prog3} \end{array} \begin{bmatrix} \text{Prog1} & \text{Prog2} & \text{Prog3} \\ 1 & 0.495 & 1 \\ 1/0.495 & 1 & 0.895 \\ 1 & 1/0.895 & 1 \end{bmatrix}$$

Now calculating geometric mean (GM) for ith row: $GM_1 = (1 \times 0.495 \times 1)1/3 = 0.7910$, $GM_2 = (1/0.495 \times 1 \times 0.895)1/3 = 1.2182$, and $GM_3 = (1 \times 1/0.895 \times 1)1/3 = 1.0376$.

Total geometric mean $= 3.0468$.

The normalised weights are: $W_1 = 0.7910/3.0468 = 0.2596$, $W_2 = 1.2182/3.0468 = 0.3998$, and $W_3 = 1.0376/3.0468 = 0.3406$.

Now consistency checking

$$\text{So the } A_3 = \begin{bmatrix} 1 & 0.495 & 1 \\ 1/0.495 & 1 & 0.895 \\ 1 & 1/0.895 & 1 \end{bmatrix} \times \begin{bmatrix} 0.2596 \\ 0.3998 \\ 0.3406 \end{bmatrix} = \begin{bmatrix} 0.7981 \\ 1.229 \\ 1.0469 \end{bmatrix}$$

$$\text{And } A_4 = \begin{bmatrix} 0.7981 \\ 1.229 \\ 1.0469 \end{bmatrix} \div \begin{bmatrix} 0.2596 \\ 0.3998 \\ 0.3406 \end{bmatrix} = \begin{bmatrix} 3.0743 \\ 3.0740 \\ 3.0736 \end{bmatrix}$$

And the maximum value λ_{max} that is the average of matrix A_4:

$$\lambda_{max} = \frac{3.0743 + 3.0740 + 3.0736}{3} = 3.073$$

$$\text{Then CI} = \frac{(\lambda_{max} - n)}{n-1} = \frac{3.073 - 3}{2} = 0.036.$$

$$\text{And CR} = \frac{CI}{RI} = \frac{0.036}{0.52} = 0.070 < 0.1.$$

Hence, the weights are consistent.

Step 4: A matrix is formed with the help of the obtained weights:

$$\begin{bmatrix} 0.2493 & 0.090 & 0.2596 \\ 0.3984 & 0.0970 & 0.3998 \\ 0.3521 & 0.8128 & 0.3406 \end{bmatrix}$$

So we can obtain the final rank:

$$\begin{bmatrix} 0.2493 & 0.090 & 0.2596 \\ 0.3984 & 0.0970 & 0.3998 \\ 0.3521 & 0.8128 & 0.3406 \end{bmatrix} \times \begin{bmatrix} 0.649 \\ 0.121 \\ 0.229 \end{bmatrix} = \begin{bmatrix} 0.2319 \\ 0.3617 \\ 0.4047 \end{bmatrix}$$

According to the higher value of the above matrix, we can decide the rank; hence, ranking is *Prog 3, Prog 2 and Prog 1*.

The FAHP is a useful methodology. Similarly, the above technique is applied in a more generalised manner by using all 17 effort multipliers of the COCOMO model considering ten hypothetical programmers as alternatives. The hierarchy of programmer ranking process based on COCOMO's effort multiplier is depicted in Figure 5.3 as there are also three layers where the upper layer represents goal and the second layer represents COCOMO's effort multipliers as 17 criteria. The last layer (leaf) represents alternatives available, i.e., the group of programmer to be ranked.

The FAHP as explained above is applied to find out the rank of the programmer. Initially, as shown in Table 5.9, we have constructed COCOMO's 17 effort multipliers as criteria and ten programmers as alternatives. After fuzzification of 17 effort multipliers, we assigned fuzzy linguistic term in each cell, with the help of human expert in this domain. The linguistic value assigned to each programmer for various criteria are conflicting in nature.

In order to apply the FAHP method, fuzzified data must be converted into precise data by applying the three-point scale of Chen and Hung (Rao, 2007) method. The fuzzy linguistic term shown in Table 5.2 is converted into numeric data as shown in Table 5.3. In the next step, we compare criteria with criteria by assigning comparative weights from Saaty's nine-point scale as shown in Table 5.4.

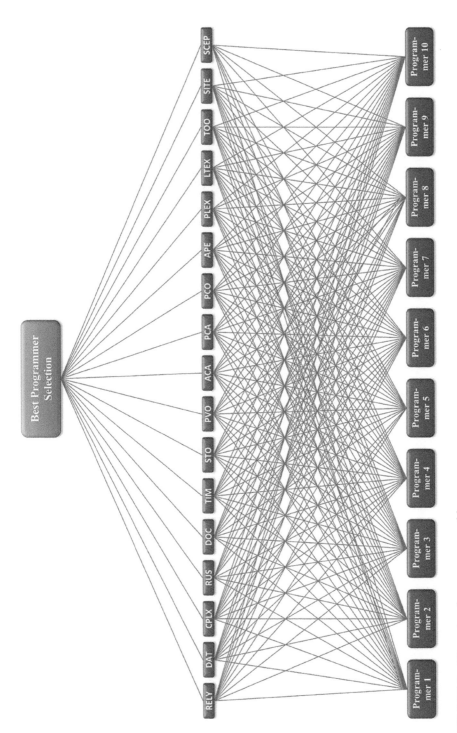

FIGURE 5.3 Hierarchy of programmer ranking.

TABLE 5.2

Decision-Making Matrix

Prgmr	RELY	DATA	CPLX	RUSE	DOCU	TIME	STOR	PVOL	ACAP	PCAP	PCON	APEX	PLEX	LTEX	TOOL	SITE	SCEP
P1	High	Average	Low	Average	High	Low	Low	High	Average	High	Low	High	Low	High	High	Average	Average
P2	Average	Low	Low	High	High	Average	Average	High	Low	High	High	High	Average	Low	Average	Average	Average
P3	Low	High	High	Average	Average	Low	High	High	High	Average	Average	Low	Average	High	Average	Average	Average
P4	High	Average	Average	High	High	Low	Low	High	Low	Average	High	High	High	High	High	Low	Average
P5	Low	Average	High	Low	High	Low	High	Average	Low	High	Average	Low	Average	Average	High	High	Average
P6	High	Average	High	Low	High	High	High	Average	Low	Average	Low	Average	Low	High	High	High	Average
P7	Average	High	Low	Average	Average	High	High	High	Low	Low	High	Average	Average	Low	Low	High	Average
P8	Average	Low	High	High	Average	Average	High	Average	Average	Low	High	Average	High	Average	Average	Low	Average
P9	Low	Average	Low	High	High	Low	High	Average	High	Average	Average	Low	Average	Average	High	Average	Average
P10	Low	High	Average	Average	High	High	Low	Average	Low	Average	Average	High	High	Low	High	Average	Average

TABLE 5.3

Fuzzy Linguistic Term into Crisp Data

Prgmr	RELY	DATA	CPLX	RUSE	DOCU	TIME	STOR	PVOL	ACAP	PCAP	PCON	APEX	PLEX	LTEX	TOOL	SITE	SCEP
P1	0.895	0.495	0.115	0.495	0.895	0.115	0.115	0.895	0.495	0.895	0.115	0.895	0.115	0.895	0.895	0.495	0.495
P2	0.495	0.115	0.115	0.895	0.895	0.495	0.495	0.895	0.115	0.895	0.895	0.895	0.495	0.115	0.495	0.495	0.495
P3	0.115	0.895	0.895	0.495	0.495	0.115	0.895	0.895	0.895	0.495	0.495	0.115	0.495	0.895	0.495	0.495	0.495
P4	0.895	0.495	0.495	0.895	0.895	0.115	0.115	0.895	0.115	0.495	0.895	0.895	0.895	0.895	0.895	0.115	0.495
P5	0.115	0.495	0.895	0.115	0.895	0.115	0.895	0.495	0.115	0.895	0.495	0.115	0.495	0.495	0.895	0.895	0.495
P6	0.895	0.495	0.895	0.115	0.895	0.895	0.895	0.495	0.115	0.495	0.115	0.495	0.115	0.895	0.895	0.895	0.495
P7	0.495	0.895	0.115	0.495	0.495	0.895	0.895	0.895	0.115	0.115	0.895	0.495	0.495	0.115	0.115	0.895	0.495
P8	0.495	0.115	0.895	0.895	0.495	0.115	0.895	0.495	0.495	0.115	0.895	0.115	0.895	0.495	0.495	0.115	0.495
P9	0.115	0.495	0.115	0.895	0.895	0.115	0.895	0.495	0.895	0.495	0.495	0.115	0.495	0.495	0.895	0.495	0.495
P10	0.115	0.895	0.495	0.495	0.895	0.895	0.115	0.495	0.115	0.495	0.495	0.895	0.895	0.115	0.895	0.495	0.495

TABLE 5.4

Relative Importance Matrix

Attributes	RELY	DATA	CPLX	RUSE	DOCU	TIME	STOR	PVOL	ACAP	PCAP	PCON	APEX	PLEX	LTEX	TOOL	SITE	SCEP
RELY	1	0.5	2	0.5	1	2	2	1	0.5	1	2	1	2	1	1	0.5	0.5
DATA	2	1	0.5	1	0.5	0.5	0.5	0.5	1	0.5	0.5	0.5	0.5	0.5	0.5	1	1
CPLX	0.5	0.2	1	0.5	2	1	1	2	0.5	2	1	2	1	2	2	0.5	0.5
RUSE	2	1	2	1	0.5	0.5	0.5	0.5	1	0.5	0.5	0.5	0.5	0.5	0.5	1	1
DOCU	1	2	0.5	2	1	2	2	1	0.5	1	2	1	2	1	1	0.5	0.5
TIME	0.5	2	1	2	0.5	1	1	2	0.5	2	1	2	1	2	2	0.5	0.5
STOR	0.5	2	1	2	0.5	1	1	2	0.5	2	1	2	1	2	2	0.5	0.5
PVOL	0.5	2	0.5	2	1	0.5	0.5	1	0.5	1	2	1	2	1	1	0.5	0.5
ACAP	2	1	2	1	2	2	2	2	1	0.5	0.5	0.5	0.5	0.5	0.5	1	1
PCAP	1	2	0.5	2	1	0.5	0.5	1	2	1	1	1	1	2	1	0.5	0.5
PCON	0.5	2	1	2	0.5	1	1	0.5	2	0.5	1	0.5	2	1	0.5	0.5	0.5
APEX	1	2	0.5	2	1	0.5	0.5	1	2	1	1	1	1	1	1	0.5	0.5
PLEX	0.5	2	1	2	0.5	1	1	0.5	2	0.5	2	2	1	2	1	0.5	0.5
LTEX	1	2	0.5	2	1	0.5	0.5	1	2	1	2	1	0.5	1	1	0.5	0.5
TOOL	1	2	0.5	2	1	0.5	0.5	1	2	1	2	1	0.5	1	1	0.5	0.5
SITE	2	1	2	1	2	2	2	2	1	2	2	s	2	2	2	1	1
SCEP	2	1	2	1	2	2	2	2	1	2	2	2	2	2	2	1	1

In step 2, geometric mean (GM = 17.2844), consistency index (CI = 0.11318456) and consistency ratio (CR = 0.072554), respectively, for checking consistency of weights are calculated. The calculated CR is less than 0.10, which shows that the weights are consistent.

In the next step, alternatives are compared with alternatives as we have done above for three criteria; we have applied the same for all the 17 criteria and presented them as pair-wise comparison matrix, and CI and CR are calculated for all matrices as above. Expert selects the shortlisted ten programmers among a group of programmers and each programmer is compared with remaining programmers based on their individual weighted values and the final ranking for best programmer is derived. In all cases, CR is in the acceptable range, which shows that our weights for all the matrices are consistent. The last step is to obtain the overall scores for the alternatives by multiplying the relative normalised weight of each attributes with normalised values for alternative, and finally, the corresponding rank of the programmers as shown in Table 5.5 are obtained. Table shows the highest value of weight for programmer $P2$; hence, $P2$ is designated as the first rank.

5.6 APPRAISAL OF PROGRAMMERS' RANK USING INTEGRATED FAHP AND FTOPSIS

TOPSIS gives a solution that is not only closest to the hypothetically best, but also the farthest from the hypothetically worst. The TOPSIS method is extended similar to FAHP with fuzzy theory and known as fuzzy TOPSIS (FTOPSIS) method. After applying fuzzy TOPSIS method using overall weights of programmers, a normalised decision matrix and a weighted normalised matrix are constructed as shown in Tables 5.6–5.8 and with the help of Excel sheet and obtained normalised values for alternatives by each criterion.

Positive ideal and negative ideal solutions with equations 5.6 and 5.7 are calculated as shown in Tables 5.9 and 5.10; also, separation measures for each alternative from positive ideal and negative ideal alternatives through equations 5.8 and 5.9 are calculated as shown in Tables 5.11 and 5.12.

TABLE 5.5
Calculated Weights and Ranks of Programmers using FAHP

Programmer ID	Weight	Rank
$P2$	0.1276	1
$P10$	0.1148	2
$P8$	0.1084	3
$P9$	0.1083	4
$P4$	0.1017	5
$P6$	0.1008	6
$P5$	0.1003	7
$P7$	0.0978	8
$P1$	0.0900	9
$P3$	0.0850	10

TABLE 5.6

Alternative Normalisation Weights for Each Criterion

Weight	0.05786	0.03848	0.05482	0.04175	0.06277	0.06277	0.06277	0.05119	0.05786	0.05786	0.0453	0.0533	0.0578	0.0533	0.0533	0.0943	0.0943
Programmer	C1	C2	C3	C4	C5	C6	C7	C8	C9	C10	C11	C12	C13	C14	C15	C16	C17
P1	0.1059	0.0690	0.1117	0.0644	0.0799	0.1045	0.1373	0.0669	0.0561	0.0796	0.0970	0.1138	0.0860	0.0978	0.0820	0.0682	0.1
P2	0.0698	0.0976	0.1117	0.0977	0.0799	0.0643	0.0597	0.0669	0.1122	0.0796	0.0770	0.1138	0.0802	0.9126	0.0666	0.0682	0.1
P3	0.0988	0.0793	0.0072	0.0740	0.0799	0.1200	0.1115	0.0669	0.1289	0.0914	0.0844	0.0862	0.0802	0.0851	0.0666	0.0682	0.1
P4	0.1059	0.0976	0.0846	0.1122	0.0918	0.1200	0.1373	0.0669	0.0977	0.0914	0.0844	0.0990	0.0922	0.0851	0.1083	0.1274	0.1
P5	0.0860	0.0911	0.0972	0.0977	0.0918	0.1200	0.0970	0.1166	0.0977	0.1050	0.0970	0.0750	0.0922	0.1048	0.1272	0.1035	0.1
P6	0.0922	0.0911	0.0972	0.0977	0.0918	0.0792	0.0970	0.1250	0.0977	0.1050	0.0970	0.1138	0.1135	0.0851	0.1272	0.1035	0.1
P7	0.1216	0.1046	0.0846	0.1122	0.1211	0.0792	0.0970	0.0883	0.0977	0.0853	0.0844	0.1138	0.1059	0.0691	0.0879	0.1035	0.1
P8	0.1216	0.1288	0.0846	0.0977	0.1211	0.1286	0.0970	0.1339	0.1289	0.0853	0.0844	0.0990	0.1059	0.1383	0.1244	0.0840	0.1
P9	0.0988	0.1202	0.0737	0.0977	0.1211	0.1045	0.0970	0.1339	0.0850	0.1385	0.1470	0.0990	0.1217	0.1383	0.1083	0.1365	0.1
P10	0.0988	0.1202	0.1693	0.1481	0.1211	0.0792	0.0686	0.1339	0.0977	0.1385	0.1470	0.0862	0.1217	0.1048	0.1010	0.1365	0.1

TABLE 5.7

Calculated Values for Normalised Matrix

Weight	0.0578	0.0384	0.0548	0.0417	0.0627	0.0627	0.0627	0.0511	0.0578	0.0578	0.0453	0.0533	0.0578	0.0533	0.0533	0.0943	0.0943
Programmer	C1	C2	C3	C4	C5	C6	C7	C8	C9	C10	C11	C12	C13	C14	C15	C16	C17
P1	0.0112	0.0047	0.0124	0.0041	0.0063	0.0109	0.0188	0.0044	0.0031	0.0063	0.0094	0.0129	0.0074	0.0095	0.0067	0.0046	0.01
P2	0.0048	0.0095	0.0124	0.0095	0.0063	0.0041	0.0035	0.0044	0.0126	0.0063	0.0059	0.0129	0.0064	0.8328	0.0044	0.0046	0.01
P3	0.0097	0.0062	0.00005	0.0054	0.0063	0.0144	0.0124	0.0044	0.0166	0.0083	0.0071	0.0074	0.0064	0.0072	0.0044	0.0046	0.01
P4	0.0112	0.0095	0.0071	0.0126	0.0084	0.0144	0.0188	0.0044	0.0095	0.0083	0.0071	0.0098	0.0085	0.0072	0.0117	0.0162	0.01
P5	0.0074	0.0083	0.0094	0.0095	0.0084	0.0144	0.0094	0.0136	0.0095	0.0110	0.0094	0.0056	0.0085	0.0109	0.0161	0.0107	0.01
P6	0.0085	0.0083	0.0094	0.0095	0.0084	0.0062	0.0094	0.0156	0.0095	0.0110	0.0094	0.0129	0.0128	0.0072	0.0161	0.0107	0.01
P7	0.0148	0.0109	0.0071	0.0126	0.0146	0.0062	0.0094	0.0078	0.0095	0.0072	0.0071	0.0129	0.0112	0.0047	0.0077	0.0107	0.01
P8	0.0148	0.0166	0.0071	0.0095	0.0146	0.0165	0.0094	0.0179	0.0166	0.0072	0.0071	0.0098	0.0112	0.0191	0.0154	0.0070	0.01
P9	0.0097	0.0144	0.0054	0.0095	0.0146	0.0109	0.0094	0.0179	0.0072	0.0192	0.0216	0.0098	0.0148	0.0191	0.0117	0.0186	0.01
P10	0.0097	0.0144	0.0286	0.0219	0.0146	0.0062	0.0047	0.0179	0.0095	0.0192	0.0216	0.0074	0.0148	0.0109	0.0102	0.0186	0.01
Sum	**0.1021**	**0.1032**	**0.0996**	**0.1045**	**0.1032**	**0.1046**	**0.1055**	**0.1088**	**0.1**	**0.1044**	**0.1059**	**0.1017**	**0.1022**	**0.9292**	**0.1049**	**0.1067**	**0.1**
Sq. Root	**0.3196**	**0.3212**	**0.3156**	**0.3233**	**0.3212**	**0.3234**	**0.3248**	**0.3299**	**0.3224**	**0.3231**	**0.32549**	**0.3190**	**0.3198**	**0.9639**	**0.3238**	**0.3267**	**0.3162**

TABLE 5.8

Calculated Values for Normalised Matrix for each alternative

Weight	0.05786	0.03848	0.05482	0.04175	0.06277	0.06277	0.06277	0.05119	0.05786	0.05786	0.0453	0.05332	0.05786	0.05332	0.05332	0.09437	0.09437
Programmer	C1	C2	C3	C4	C5	C6	C7	C8	C9	C10	C11	C12	C13	C14	C15	C16	C17
P1	0.3314	0.2149	0.3540	0.1994	0.2488	0.3231	0.4226	0.2030	0.1740	0.2463	0.2980	0.3567	0.2690	0.1014	0.2534	0.2090	0.3162
P2	0.2186	0.3039	0.3540	0.3022	0.2488	0.1989	0.1839	0.2030	0.3481	0.2463	0.2365	0.3567	0.2510	0.9467	0.2058	0.2090	0.3162
P3	0.3092	0.2469	0.0229	0.2290	0.2488	0.3712	0.3432	0.2030	0.3998	0.2829	0.2594	0.2703	0.2510	0.0883	0.2058	0.2090	0.3162
P4	0.3314	0.3039	0.2683	0.3471	0.2858	0.3712	0.4226	0.2030	0.3030	0.2829	0.2594	0.3105	0.2883	0.0883	0.3344	0.3900	0.3162
P5	0.2692	0.2836	0.3082	0.3022	0.2858	0.3712	0.2988	0.3535	0.3030	0.3250	0.2980	0.2353	0.2883	0.1087	0.3928	0.3168	0.3162
P6	0.2885	0.2836	0.3082	0.3022	0.2858	0.2449	0.2988	0.3789	0.3030	0.3250	0.2980	0.3567	0.3550	0.0883	0.3928	0.3168	0.3162
P7	0.3807	0.3257	0.2683	0.3471	0.3771	0.2449	0.2988	0.2679	0.3030	0.2640	0.2594	0.3567	0.3312	0.0717	0.2716	0.3168	0.3162
P8	0.3807	0.4011	0.2683	0.3022	0.3771	0.3978	0.2988	0.4060	0.3998	0.2640	0.2594	0.3105	0.3312	0.1434	0.3841	0.2573	0.3162
P9	0.3092	0.3742	0.2336	0.3022	0.3771	0.3231	0.2988	0.4060	0.2638	0.4288	0.4517	0.3105	0.3805	0.1434	0.3344	0.4180	0.3162
P10	0.3092	0.3742	0.5367	0.4581	0.3771	0.2449	0.2113	0.4060	0.3030	0.4288	0.4517	0.2703	0.3805	0.1087	0.3120	0.4180	0.316

TABLE 5.9
Weighted Normalised Decision Matrix for Alternative

Weight	0.05786	0.03848	0.05482	0.04175	0.06277	0.06277	0.06277	0.05119	0.05786	0.05786	0.0453	0.05332	0.05786	0.05332	0.05332	0.09437	0.09437
Programmer	C1	C2	C3	C4	C5	C6	C7	C8	C9	C10	C11	C12	C13	C14	C15	C16	C17
P1	0.0191	0.0082	0.0194	0.0083	0.0156	0.0202	0.0265	0.0103	0.0100	0.0142	0.0135	0.0190	0.0155	0.0054	0.0135	0.0197	0.0298
P2	0.0126	0.0116	0.0194	0.0126	0.0156	0.0124	0.0115	0.0103	0.0201	0.0142	0.0107	0.0190	0.0145	0.0504	0.0109	0.0197	0.0298
P3	0.0178	0.0095	0.0012	0.0095	0.0156	0.0233	0.0215	0.0103	0.0231	0.0163	0.0117	0.0144	0.0145	0.0047	0.0109	0.0197	0.0298
P4	0.0191	0.0116	0.0147	0.0144	0.0179	0.0233	0.0265	0.0103	0.0175	0.0163	0.0117	0.0165	0.0166	0.0047	0.0178	0.0368	0.0298
P5	0.0155	0.0109	0.0168	0.0126	0.0179	0.0233	0.0187	0.0180	0.0175	0.0188	0.0135	0.0125	0.0166	0.0057	0.0209	0.0299	0.0298
P6	0.0166	0.0109	0.0168	0.0126	0.0179	0.0153	0.0187	0.0193	0.0175	0.0188	0.0135	0.0190	0.0205	0.0047	0.0209	0.0299	0.0298
P7	0.0220	0.0125	0.0147	0.0144	0.0236	0.0153	0.0187	0.0137	0.0175	0.0152	0.0117	0.0190	0.0191	0.0038	0.0144	0.0299	0.0298
P8	0.0220	0.0154	0.0147	0.0126	0.0236	0.0249	0.0187	0.0207	0.0231	0.0152	0.0117	0.0165	0.0191	0.0076	0.0204	0.0242	0.0298
P9	0.0178	0.0144	0.0128	0.0126	0.0236	0.0202	0.0187	0.0207	0.0152	0.0248	0.0204	0.0165	0.0220	0.0076	0.0178	0.0394	0.0298
P10	0.0178	0.0144	0.0294	0.0191	0.0236	0.0153	0.0132	0.0207	0.0175	0.0248	0.0204	0.0144	0.0220	0.0057	0.0166	0.0394	0.0298

TABLE 5.10
Determine the Positive Ideal and Negative Ideal Solutions

	C1	C2	C3	C4	C5	C6	C7	C8	C9	C10	C11	C12	C13	C14	C15	C16	C17
v_j^*	0.0220	0.01543	0.0294	0.0191	0.0236	0.0249	0.0265	0.0207	0.0231	0.02481	0.02046	0.0190	0.02201	0.0504	0.020945	0.0394	0
v_j'	0.0126	0.0082	0.0012	0.0083	0.0156	0.0124	0.0115	0.0103	0.0100	0.01425	0.0107	0.01255	0.01452	0.0038	0.0109	0.0197	0

TABLE 5.11

Positive Ideal Values for Alternatives

Programmer	C1	C2	C3	C4	C5	C6	C7	C8
P1	8.13074E-06	5.131E-05	0.0001	0.000116669	6.5E-05	2.19822E-05	0	0.0001
P2	8.7915E-05	1.397E-05	0.0001	4.23513E-05	6.5E-05	0.00015591	0.0002244	0.0001
P3	1.71041E-05	3.521E-05	0.0007	9.14583E-05	6.5E-05	2.79653E-06	2.48E-05	0.0001
P4	8.13074E-06	1.397E-05	0.0002	2.145E-05	3.3E-05	2.79653E-06	0	0.0001
P5	4.16271E-05	2.044E-05	0.0001	4.23513E-05	3.3E-05	2.79653E-06	6.037E-05	7.2401E-06
P6	2.84496E-05	2.044E-05	0.0001	4.23513E-05	3.3E-05	9.21634E-05	6.037E-05	1.9372E-06
P7	0	8.397E-06	0.0002	2.145E-05	0	9.21634E-05	6.037E-05	5.0027E-05
P8	0	0	0.0002	4.23513E-05	0	0	6.037E-05	0
P9	1.71041E-05	1.068E-06	0.0002	4.23513E-05	0	2.19822E-05	6.037E-05	0
P10	1.71041E-05	1.068E-06	0	0	0	9.21634E-05	0.0001759	0

TABLE 5.12

Negatives Ideal Values for Alternatives

Programmer	C1	C2	C3	C4	C5	C6	C7	C8
P1	4.25737E-05	0	0.00032947	0	0	6.0807E-05	0.0002244	0
P2	0	1.174E-05	0.00032947	1.84342E-05	0	0	0	0
P3	2.74638E-05	1.513E-06	0	1.53244E-06	0	0.0001	0.0001	0
P4	4.25737E-05	1.174E-05	0.00018093	3.80675E-05	5.4E-06	0.0001	0.0002244	0
P5	8.55207E-06	6.984E-06	0.00024457	1.84342E-05	5.4E-06	0.0001	5.2E-05	5.934E-05
P6	1.63417E-05	6.984E-06	0.00024457	1.84342E-05	5.4E-06	8.33029E-06	5.2E-05	8.1039E-05
P7	8.7915E-05	1.82E-05	0.00018093	3.80675E-05	6.5E-05	8.33029E-06	5.2E-05	1.1029E-05
P8	8.7915E-05	5.131E-05	0.00018093	1.84342E-05	6.5E-05	0.0001	5.2E-05	0.00010804
P9	2.74638E-05	3.757E-05	0.00013333	1.84342E-05	6.5E-05	6.0807E-05	5.2E-05	0.00010804
P10	2.74638E-05	3.757E-05	0.00079311	0.000116668	6.5E-05	8.33029E-06	2.95E-06	0.00010804

C9	C10	C11	C12	C13	C14	C15	C16	C17	SUM	SQ. ROOT (S*i)
0.000171	0.0001	4.85E-05	0	4.159E-05	0.0020	5.523E-05	0.00039	0.0008	0.00421	0.0648
8.97E-06	0.0001	9.5E-05	0	5.612E-05	0	9.937E-05	0.00039	0.0008	0.00245	0.0494
0	7.1286E-05	7.59E-05	2.122E-05	5.612E-05	0.0020	9.937E-05	0.00039	0.0008	0.00484	0.0695
3.14E-05	7.1286E-05	7.59E-05	6.065E-06	2.842E-05	0.0020	9.695E-06	7E-06	0.0008	0.00362	0.0601
3.14E-05	3.6105E-05	4.85E-05	4.189E-05	2.842E-05	0.0019	0	9.1E-05	0.0008	0.00353	0.0594
3.14E-05	3.6105E-05	4.85E-05	0	2.174E-06	0.0020	0	9.1E-05	0.0008	0.00363	0.0602
3.14E-05	9.0999E-05	7.59E-05	0	8.124E-06	0.00217	4.175E-05	9.1E-05	0.0008	0.00386	0.0620
0	9.0999E-05	7.59E-05	6.065E-06	8.124E-06	0.0018	2.136E-07	0.00023	0.0008	0.00346	0.0587
6.2E-05	0	0	6.065E-06	0	0.0018	9.695E-06	0	0.0008	0.00322	0.0567
3.14E-05	0	0	2.122E-05	0	0.0019	1.856E-05	0	0.0008	0.00324	0.0569

C9	C10	C11	C12	C13	C14	C15	C16	C17	SUM	SQ. ROOT (S'i)	S*i+S'i
0	0	7.76E-06	4.188E-05	1.087E-06	2.51E-06	6.437E-06	0	0.0008	0.0016	0.0400	0.1049
0.0001	0	0	4.188E-05	0	0.0021766	0	0	0.0008	0.0035	0.0597	0.1092
0.000171	4.4914E-06	1.08E-06	3.48E-06	0	7.814E-07	0	0	0.0008	0.0013	0.0363	0.1058
5.57E-05	4.4914E-06	1.08E-06	1.607E-05	4.666E-06	7.814E-07	4.699E-05	0.00029	0.0008	0.0019	0.0439	0.1041
5.57E-05	2.0736E-05	7.76E-06	0	4.666E-06	3.891E-06	9.937E-05	0.0001	0.0008	0.0017	0.0412	0.1006
5.57E-05	2.0736E-05	7.76E-06	4.188E-05	3.62E-05	7.814E-07	9.937E-05	0.0001	0.0008	0.0016	0.0411	0.1013
5.57E-05	1.0466E-06	1.08E-06	4.188E-05	2.154E-05	0	1.23E-05	0.0001	0.0008	0.0015	0.0398	0.1019
0.000171	1.0466E-06	1.08E-06	1.607E-05	2.154E-05	1.463E-05	9.037E-05	2.1E-05	0.0008	0.0019	0.0441	0.1028
2.7E-05	0.00011156	9.5E-05	1.607E-05	5.612E-05	1.463E-05	4.699E-05	0.00039	0.0008	0.0021	0.0463	0.1031
5.57E-05	0.00011156	9.5E-05	3.48E-06	5.612E-05	3.891E-06	3.204E-05	0.00039	0.000891	0.0028	0.0528	0.1098

At last, we have calculated the relative closeness to the ideal solution C_i^* using equation 5.10 and the corresponding ranks of the programmers as shown below in Table 5.13 are found.

5.7 COMPARATIVE ANALYSIS

To find out the rank of programmers, two MCDM methods are applied. Figure 5.4 shows the comparative results of two MCDM methods; the ranks obtained through these techniques are, however, different in case of fuzzy AHP and fuzzy TOPSIS methods, but both the techniques have designated $P2$ as the first rank. Ranks of other programmers are also very close to each other.

TABLE 5.13
Calculated Weights and Ranks of Programmers Using FTOPSIS

Programmer ID	Weight	Rank
P_2	0.54700	1
P_{10}	0.48143	2
P_9	0.44959	3
P_8	0.42874	4
P_4	0.42220	5
P_5	0.40958	6
P_6	0.40554	7
P_7	0.39098	8
P_1	0.38193	9
P_3	0.34304	10

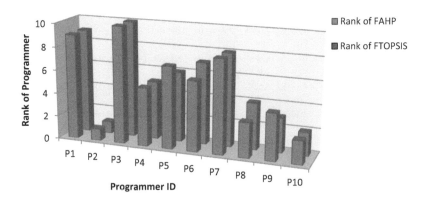

FIGURE 5.4 Comparative graph of ranking using FAHP and FTOPSIS.

5.8   CONCLUSIONS

Software effort estimation is highly uncertain and ambiguous; therefore, fuzzy logic-based MCDM methods may be well suited. COCOMO's effort estimation method is reliable and widely used, which is based on 17 effort multipliers. In order to select a group of programmers for better software effort estimation, fuzzy versions of AHP and TOPSIS are utilised and ranks of programmers based on sample data collected are evaluated. Ranks found in case of these techniques are different, but both the techniques have produced the same rank (Rank 1) to programmer P2. Further, this process can be repeated for larger group of programmers using other MCDM methods.

REFERENCES

1. Pogarcic, I., Francic, M. and Davidovic, V. (2008), *Application of AHP Method in Traffic Planning*, ISEP.
2. Mishra, S. K. and Ray, A. (2012), Software developer selection: a Holistic approach for an eclectic decision, *International Journal of Computer Application* 47, 1, 12–18.
3. Yuen, K. K. F. and Lau, H. C.W. (2008), Software vendor selection using fuzzy analytic hierarchy process with ISO/IEC 9126, *IAENG International Journal of Computer Science* 35, 3.
4. Gungor, Z., Serhadlıoglu, G. and Kesen, S. E. (2009), A fuzzy AHP approach to personnel selection problem, *Applied Soft Computing* 9, 641–646.
5. Cebeci, U. (2009), Fuzzy AHP-based decision support system for selecting ERP systems in textile industry by using balanced scorecard, *Expert Systems with Applications* 36, 8900–8909.
6. Buyukozkan, G., Cifci, G. and Guleryuz, S. (2011), Strategic analysis of healthcare service quality using fuzzy AHP methodology, *Expert Systems with Applications* 38, 9407–9424.
7. Catak, F.O., Karabas, S. and Yildirim, S. (2012), Fuzzy analytic hierarchy based DBMS selection in Turkish national identity card management project, *International Journal of Information Sciences and Techniques (IJIST)* 2, 4, 29–38.
8. Javanbarg, M. B., Scawthorn, C., Kiyono, J. and Shahbodaghkhan, B. (2012), Fuzzy AHP-based multi-criteria decision making systems using particle swarm optimization, *Expert Systems with Applications* 39, 960–966.
9. Kabir, G. and Hasin, M. A. A. (2011), Comparative analysis of AHP and fuzzy AHP models for multicriteria inventory classification, *International Journal of Fuzzy Logic Systems* 1, 1, 1–16.
10. Sehra, S. K., Brar, Y. S. and Kaur, N. (2012), Multi criteria decision making approach for selecting effort estimation model, *International Journal of Computer Applications (0975–8887)* 39, 1, 10–17.
11. Bondor, C. I. and Muresan, A. (2012), Correlated criteria in decision models: Recurrent application of TOPSIS method, *Applied Medical Informatics* 30, 1, 55–63.
12. Wimatsari, G. A. M. S., Putra, K. G. D. and Buana, P.W. (2013), Multi-attribute decision making scholarship selection using a modified fuzzy TOPSIS, *IJCSI International Journal of Computer Science Issues* 10, 1 and 2, 309–317.
13. Pressman, R.S. (2005), *Software Engineering*, McGraw Hill Higher Education.
14. Hota, H.S., Singhai, S.K., Shukla, R. (2012), Application of fuzzy analytic hierarchy method in software engineering scenario, *International Journal of Computer Applications (0975–8887)* 57, 21, 45–50.
15. Milicic, D. (2004), *Applying COCOMO II-A Case Study, Master Thesis Software Engineering*, Thesis no: MSE-2004-19.

16. Zhu, K. J., Jing, Y., and Chang, D. Y. (1999), A discussion on extent analysis method and applications of fuzzy-AHP, *European Journal of Operational Research*, 116, 450–456.
17. Rao, R. V. (2007), *Decision Making in the Manufacturing Environment*, Springer-Verlag London Limited.
18. Saaty, T. L. (1980), *The Analytical Hierarchy Process*, McGraw Hill, New York.

6 Implementing Multi-Criteria Decision-Making to Detect Potential Onset of Heart Disease

Narina Thakur
Bhagwan Parshuram Institute of Technology

Sardar M. N. Islam
ISILC, Victoria University

Isha Bansal, Aakriti and Kartik Gupta
Bharati Vidyapeeth's College of Engineering

Rachna Jain
Bhagwan Parshuram Institute of Technology

CONTENTS

DOI: 10.1201/9780367816414-6

6.1   INTRODUCTION

Cardiovascular diseases (CVDs) are the leading cause of death, killing an estimated 17.9 million people each year [1]. CVDs are a group of heart and vascular diseases, including coronary artery disease, rheumatic heart disease and cerebrovascular disease. Heart attacks and strokes account for more than four from every five CVD deaths, and one-third of these deaths occur in people under the age of 70. Heart disease is the main cause of death in India. The absolute estimated prevalence of CVD in India is 54.6 million. The deaths because of suffering from heart disease have decreased by approx. 39% from 2001 to 2011. In fact, many concerns about the nutrition of our teenagers seem to be related to the early phase of cardiovascular disease in the arteries or the thickening of the coronary arterial walls [2]. This is a steady, silent disease that typically progresses for decades before anyone shows symptoms. So, the need to go to the doctor often is observed after the sixth or seventh decade. This is why there is a need to screen patients who may develop CVD in the near future based on their present lifestyle, and they should know they must take care of their health to prevent them from ending up in a hospitalisation. Several risk factors are relevant to decide whether you are likely to develop cardiovascular disease. Few parameters such as age and inheritance are not in the control of the individual, and few activities such as eating habits, fitness, exercise and lifestyle can be changeable or controllable. The risk of suffering from heart disease increases at the age of 55 years in women and 45 years in men [3]. If you have close relatives with a history of heart disease, your risk may be higher. Many risk factors include obesity, insulin or diabetes, elevated cholesterol and blood pressure, a family history of heart failure/diseases, inactivity, a poor diet, smoking and clinical depression. There are other risk factors for heart disease, but genetic factors can increase the risk of developing heart disease, and unhealthy lifestyle and personal choices also play an important role. Several unhealthy habits that may lead to heart disease include unhealthy lifestyles with not enough physical activity, a poor nutrition diet. Life's Simple 7s to lead a healthy lifestyle include not smoking, physical activity, a balanced diet, maintaining body weight, and controlling cholesterol, blood pressure and blood sugar. Hence, a stress-free and diabetes-free atmosphere is necessary. So, it is extremely important to recognise the odds of an individual having heart disease or not. According to the WHO [4], 17.9 million deaths from CVD are expected to occur annually, approximately 32% of all fatalities worldwide. The estimate is based on the most recent analysis and results of CVD deaths. Deaths by heart attacks and strokes account for more than 85% of all deaths. Heart attacks and strokes account for more than four out of every five CVD deaths, with premature deaths accounting for one-third of these deaths in those under the age of 70.

In recent years, many authors have done a significant amount of research on predicting heart attacks using a variety of techniques and algorithms. Researchers have explored the techniques based on various fields such as deep learning, machine learning and data mining. Everyone aims at improving accuracy, getting more and more accurate results. MCDM techniques have effectively been applied to prediction in numerous fields, including supportable vitality management, energy planning, transportation, geographical data systems, budgeting and asset designation. The objective of this chapter is to propose an accurate predictive model using the multi-criteria decision-making (MCDM) algorithm for the Heart Disease UCI dataset. This chapter proposes a prediction model for the risk of heart disease based on various factors and features for the Heart Disease UCI dataset by utilising the MCDM. The MCDM-based heart disease prediction model predicts if a person is at risk of developing cardiac disease in a much easier way. The model employs the dataset that constitutes 1000 individuals with a history of heart disease, and it predicts the probability of an individual to suffer from heart disease by employing MCDM and normalisation techniques. This chapter is divided into various sections as follows: Section 6.2 details the various studies and research work that has already been undertaken and related to the prediction of heart disease. Section 6.3 describes the methodology and Section 6.4 discusses the results and analyses, followed by the Conclusion section of the research.

6.2 LITERATURE REVIEW

Nason et al. [5] elaborated on cardiovascular disease prediction techniques employed in recent research and summarised their strengths and weaknesses. The paper highlights the key background issues that need to be involved in the research study. The authors presented a research study focusing on four countries: Australia, the United Kingdom, Canada, and New Zealand. The research elucidates the current state of CVD, CVD research, and the context for case studies of specific CVD research studies. The Payback Framework, which has been utilised by the UK study team in prior health research investigations, was discussed in the case studies. The medical industry is a massive reserve of relevant information. Hence, the availability of such data becomes of key importance so that valid information relevant to us can be extracted. This huge amount of data is critical for retrieving meaningful information and generating correlations among features. Chadha et al. [6] published in-depth analyses of cardiac disease prediction using data mining techniques. The primary aim of the research is to consolidate, summarise and assess various data mining strategies for heart disease prediction, which have been proposed and deployed in recent years. The neural network (NN) has been found to be more powerful and better than the other methods such as the naive Bayes (NB) model and decision tree (DT) model.

Abdul-Aziz et al. in 2019 [7] highlighted that heart disease strikes with much ferocity and medical data are still statistics and knowledge deficient. As a consequence, an essential task for medical support is appropriately diagnosing patients in a timely manner. A hospital's incorrect diagnosis results in a loss of reputation. The most important biomedical issue is the correct diagnosis of heart disease. The primary objective of the research was to use data mining techniques to provide an effective remedy for restorative circumstances. Later, Salman et al. [8] suggested that some particular medical rules revolving around chronic heart disease must be followed in cases of triage, and their urgency should be ranked based on an individual's vital signs and their attributes. The purpose of the study was to measure and evaluate vast amounts of data from patients suffering with chronic heart disease and those who require immediate intervention. A practical learning study was conducted on 500 patients with chronic heart disease who had varying symptoms and were in various stages of emergency. The paper concluded that multi-attribute and multi-criteria decision-making can assist researchers studying patients with heart problems in dealing with the challenge of storing and using vast volumes of data. New methodologies born from the study lay the foundation and improve the decision-making process in triage and effective assessment of these patient types. Raju et al. [9] analysed data mining classification techniques such as DTs, NNs, Bayesian classifiers, support vector machines (SVMs), association rule and k-nearest neighbour (k-NN) classification and employed them to diagnose cardiac diseases. The SVM is the most accurate of these techniques.

Marimuthu et al. [10] proposed a machine learning-based prediction model for heart disease by considering blood pressure, hypertension, diabetes, the numbers of cigarettes smoked each day and medical input information as input. The experiments have been performed on k-NN, NB, SVM and DT prediction models. The results demonstrated that all the proposed models appreciably accurately predict the overall risk of heart disease. Singh et al. [11] proposed that the advanced data mining techniques are effective to provide relevant results to make smart data decisions to overcome the issue of extraction of hidden information from massive volumes of data. It can help decision-makers to make better decisions. In the study, a NN is used to develop an effective heart disease prediction system (EHDPS) for detecting the risk level of heart disease. Age, gender, blood pressure, cholesterol and obesity were among the 15 medical parameters employed in the system to make accurate predictions. The EHDPS allows for the establishment of substantial knowledge, such as correlations between medical parameters linked to heart disease prediction and heart disease patterns. A multilayer perceptron neural network with backpropagation was utilised for training, yielding efficient outcomes. A fuzzy rule-based technique was proposed in [12] with a DT for predicting heart disease diagnosis. The results obtained concluded the proposed technique has an accuracy of 88%, which is statistically remarkable for diagnosing patients with cardiac disease and surpasses some existing techniques. It is unfortunate that the ever-increasing sources of information generated by hospital patient records, including records of valuable medical research resources, are not properly mined. Currently, these data are primarily used for therapeutic purposes only. These data are often used to better understand the hidden patterns and associations that can lead to better diagnosis, medicine and treatment, as well as a platform for better understanding the mechanisms driving practically every aspect of the medical realm. However, the finding of these hidden correlations is usually ignored.

Mehmood et al. [13] proposed CardioHelp, which employs convolutional neural network (CNN) deep learning method to predict the risk of a patient having heart disease. The methodology is concerned with temporal data modelling and employs CNN for early heart function prediction. The heart disease dataset was created, and the results were compared to existing methodologies, yielding positive results. Experimental results show that the CardioHelp obtains 97% accuracy and is superior to existing methods in terms of performance evaluation tools. Isola et al. [14] proposed that the massive data store can be used to make a better diagnosis based on historical data. The medical data can be efficiently mined by combining neural networks, k-NN, storage and acquisition of large-capacity memory to improve the accuracy of diagnosis

The main points of differential diagnosis are the probability of occurrence of a specific disease, which can be obtained from medical data. The system is based on a service-oriented architecture that includes diagnostics, information portals and other services. This algorithm can be used to solve some of the more common auto-discovery problems these days: diagnosis of multiple illnesses with multiple symptoms. Ayon et al. [15] presented a comparative analysis performed using seven computational intelligence techniques: logistic regression (LR), SVM, deep neural network (DNN), DT, NB, random forest (RF) and k-NN. Statlog and Cleveland heart disease datasets downloaded from the UCI machine learning database were used to measure the performance of each technique using multiple scoring techniques. In the study, a DNN achieved an accuracy of over 98.15%. Hassani et al. [16] presented a novel approach with the goal of discovering a significant method for predicting heart disease. The Cleveland dataset and the Statlog heart disease datasets from the UCI ML repository were employed to generate a unique dataset for the research. The new data include 568 cases and 14 medical parameters such as age, gender and blood pressure for heart disease training and prediction. The paper proposed a novel neural network and decision tree approach for improved cardiac disease prediction, which utilises a NN for training and a DT for testing classification. The proposed approach was compared to the NB, SVM, NN, voted perceptron and DT algorithms in terms of performance. The findings revealed that the accuracy and precision were both improved. Devansh Shah et al. [17] presented various attributes related to heart disease and proposed a model for the heart disease patients dataset from the Cleveland database of the UCI repository using supervised learning algorithms such as NB, DT, k-NN and random forest (RF) algorithms to predict the likelihood of patients having heart disease. Only 14 features are tested out of 76 attributes in the dataset; however, they are critical in establishing the performance of different algorithms. The k-NN has the highest accuracy score, according to the data.

Nagaprasad et al. [18] proposed a hybrid method by exploiting the backpropagation method in combination with the k-means clustering method, to cluster knowledge to make an improved prediction performance for the cardiac disease data sample collected from the UCI repository comprising the output of the implemented algorithm. There are 66 attributes in the sample. Every research, however, requires a subgroup of 14 criteria. Machine learning research uses the Cleveland platform. The study was designed according to current methods, accuracy, error detection and deployment time (using numerical averaging). Gavhane et al. [19] emphasised the need for effective mechanisms to recognise the symptoms of a heart attack early and avoid the development of heart attacks in children and adolescents. It isn't practical for the average person to undergo expensive tests such as ECG on a routine basis, so a convenient and reliable system must be in place to predict the risk of cardiac disease. The authors of the research developed an app that can predict one's susceptibility to heart disease based on basic indicators such as age, gender and heart rate. Neural network machine learning algorithm was adopted because it has been proven to be the most accurate and reliable algorithm. Farzana Tasnim et al. [20] analysed various data mining classification techniques, including NB, SVM, k-NN, DT, NN, LR, RF and gradient boosting for predicting the probability of heart disease using the cardiac disease dataset from the UCI machine learning repository. Traditional machine learning algorithms perform better with the feature selection strategy. The RF method with PCA has the best accuracy of 92.85% among the other algorithms. Islam et al. [21] proposed a PCA-based hybrid heuristic model to detect CHD by employing the hybrid genetic algorithm (HGA) with k-means used for final clustering. Early heart disease can be predicted with an accuracy of 94.06%.

Vafaei et al. [22] evaluated MCDM and normalisation techniques. They focused on six well-known normalisation techniques for usage in the TOPSIS method. Ma et al. [23] studied MCDM problems that can help solve and handle the cognitive limitations that can occur in many problems in real world. Lepri et al. [24] presented an overview of available technical methods to refine algorithmic decision-making fairness, accountability and transparency in the work. They also stressed the importance of bringing together multidisciplinary teams of researchers, practitioners, policy makers and citizens to jointly develop, deploy and evaluate algorithmic decision-making procedures that optimise fairness and transparency in real-world settings. Vafaei et al. [25] discussed that normalisation plays an essential role in any decision-making algorithm. The purpose of the article was to find a suitable normalisation technique that enables data fusion, which has become difficult with the advent of cyber-physical systems. The purpose of this paper is to examine metrics to determine which normalisation procedures are most appropriate for decision problems, especially the multi-criteria analytic hierarchy process (AHP) method. The researchers illustrated the relation among cyber systems. The researchers' goal was to find the best normalisation method for the AHP method.

Lakshmi et al. [26] studied TOPSIS and applied various normalisation techniques to achieve the optimal solution. The researchers concluded that linear sum-based normalisation is the best method for both time and space. Chowdhury et al. [27] observed that both single and integrated MCDM methods could be used in this area. The authors proposed future research directions, including reviewing and formulating strategies for specific CS initiatives. Asadabadi et al. [28] applied foundational MCDM methods proposed by Saaty, in particular AHP and analytic network process (ANP). The paper validates the application of MCDM instead of traditional approaches when ranking individuals for CVD prediction. The study accepted a generic company's point of view that they might not find it different or useful, but still hold their ground in the benefits of their methods. It accepts that individuals may be ranked using AHP, which a rational person might not even consider.

Adunlin et al. [29] discussed the systematic review trend analysis when MCDA is applied in the sector of health care. A total of 66 citations met the selection criteria. The increase in publishing trend occurred in the years 1990, 1997, 1999, 2005, 2008 and 2012. This trend indicated that the number of releases peaked in 2012. Frazão et al. [30] were to frame and recreate articles found in the literature, clubbing MCDA and health care together, and to evaluate common and methodological issues, compiling them into a single framework. It may include studies aimed at methodological applications of MCDA without using mathematical methods. The studies included in the paper only focused on descriptive research with no mathematical formulae derived from the texts. This pointed the paper only in the direction of MCDA's methodological application. The TOPSIS method's two major deficiencies are the non-meaningfulness of the resulting rankings in mixed data contexts and rank reversals or ranking irregularities. A meaningful mixed data TOPSIS method (TOPSIS-MMD) was proposed in [31]. The TOPSIS method was enhanced by extending the mixed data in a comparatively defensible manner.

6.3 METHODOLOGY

The key approach adopted in this paper was by analysis of medicine, machine learning, computer science and engineering journals and publications. This paper attempts to foresee the danger of coronary illness in an individual relying upon the features provided in the dataset on applying MCDM. The study is quantitative since it deals with the statistics of the various features of a person to recognise the risk probability of the person to have cardiovascular disease. The dataset used is a secondary dataset taken up from Kaggle titled "Heart Disease UCI" provided by Ronit on the platform, which provides the experiments with the Cleveland database, which simply attempt to differentiate between the existence and lack of cardiac conditions. The MCDM is a decision-making analysis that uses scores and weights as a reference to evaluate multiple criteria as part of the decision-making process and is an open and explicit cost-benefit analysis. It provides insight into different judgements of value when compared to other methods.

6.3.1 Multi-Criteria Decision-Making (MCDM) Algorithm

The MCDM is a method used for making decisions, while more than one criterion need to be taken into consideration collectively to rank or select from the evaluated choices. The MCDM involves a number of things, including choosing the attributes to be taken into consideration and then evaluating and comparing them. It also involves assigning weights representing the importance of the attributes and, in the end, making effective decisions. The MCDM is designed to reduce the occurrence and impact of "intuitive" decision-maker bias and group decision-making errors that almost inevitably undermine intuitive choices. The MCDM leads to more transparent and consistent decisions by expressing weights and associated switching between standards in a structured way. It gives us the result more sensible and with the extra real facts that pop out of it.

All the steps involved in the MCDM algorithm are mentioned in Figure 6.1. Each step is discussed in the coming sections.

6.3.1.1 Categorisation of Features

The features are categorised as beneficial and non-beneficial features, where the beneficial features are the ones whose higher values are desired, such as efficiency or profit, while non-beneficial features are the features whose lower values are required, such as cost. For instance, a better television would be the one with low cost and high picture quality. Hence, cost is a non-beneficial feature and picture quality is a beneficial feature.

6.3.1.2 Normalisation of Data

Each criterion can be calculated at different units, such as degrees, kilograms or meters. However, all these must be standardised to achieve dimensional classifications, namely a common numerical range/scale, so that the aggregation is made possible for the final score. The first step is normalisation in the decision-making process

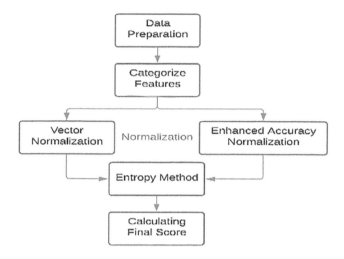

FIGURE 6.1   MCDM algorithm.

to convert data into nearly equivalent units on a common scale. Normalisation is an important step that helps improve model stability, boosts up the training procedure, and also helps in giving "equal" considerations for each feature of the data. Data standardisation is also a key aspect of the decision-making process because it makes input data quantified and compared. The data are normalised so that all the data can be compared to each other.

Normalisation intends to scale a variable somewhere in the range of 0 and 1. There are five common normalisation techniques [32] such as vector normalisation, linear min-max normalisation, linear sum-based normalisation, linear max normalisation and Gaussian normalisation. The min-max normalisation is one of the most frequent methods of data normalisation. One of the important disadvantages of the min-max standardisation is that the outliers are not well treated in this normalisation. If 99 values range from 0 to 40 and one value is 100, then the 99 values are all converted to values ranging from 0 to 0.4. The linear sum normalisation method sums the scores for each criterion and divides the score for each feature.

6.3.1.3   Vector Normalisation

We apply vector normalisation for beneficial and non-beneficial features separately. The reason why vector normalisation was chosen was that there were many 0 values in our data.

For non-beneficial features, we calculate the new value of each data item by equation 6.1.

$$\bar{x}_{ij} = \frac{x_{ij}}{\sqrt{\sum x_{ij}^2}} \tag{6.1}$$

where x_{ij} denotes the current data value and $\sqrt{\sum x_{ij}^2}$ denotes the square root of the sum of the squares of all data values of the particular column.

For beneficial features, we have equation 6.2, where each data value is changed according to the equation.

$$\overline{x}_{ij} = 1 - \left(\frac{x_{ij}}{\sqrt{\sum x_{ij}^2}} \right) \tag{6.2}$$

where x_{ij} denotes the current data value and $\sum x_{ij}^2$ denotes the square root of the sum of the squares of all data values of the particular column.

6.3.1.4   Enhanced Accuracy Normalisation

The second normalisation method applied on the data was enhanced accuracy normalisation of data. For beneficial features, we calculate the new value for each data item by equation 6.3.

$$\overline{x}_{ij} = 1 - \frac{x_{max}^j - x_{ij}}{\sum_{i=1}^{m} x_{max}^j - x_{ij}} \tag{6.3}$$

For non-beneficial features, we calculate the new value for each data item by equation 6.4.

$$\overline{x}_{ij} = 1 - \frac{x_{ij} - x_{min}^j}{\sum_{i=1}^{m} x_{ij} - x_{min}^j} \tag{6.4}$$

6.3.1.5   Entropy Method to Assign Weightage

After the normalisation is done successfully, the features in the normalised decision matrix are given weightage, which implies the importance of every feature. Now for the determination of weights of the features, the entropy method was used, which can be done using the following steps and equations shown below.

The entropy of each feature individually is calculated with the help of equation 6.5 mentioned below:

$$E_i = -h \sum_{i=1}^{m} x_{ij} \log(x_{ii}) \tag{6.5}$$

where E_i denotes the entropy of the current feature and x_{ii} denotes the current data value in the normalised data matrix. m denotes the total number of data points. And h is given by equation 6.6.

$$h = \frac{1}{(\log(m))} \tag{6.6}$$

After the calculation of entropy of each feature, the weight vector is calculated for each feature, which is taken as the objective weightage of the feature. Then, the weight vector is calculated using equation 6.7.

$$W_i = \frac{1 - E_i}{\Sigma\left(1 - E_i \right)} \tag{6.7}$$

where W_i denotes the weight vector of the particular feature, E denotes the entropy of a particular feature, and $1 - E$ denotes the degree of diversification.

The entropy method has extensively been practised as a significant model for weight determination. It is always accurate and useful, but the results are prone to distortion when too many zero values are encountered. On calculating the weight vector for each feature, the next step that comes into the picture is the multiplication of each data value of a feature with the weightage vector of that particular feature. And as a result, a weightage normalised decision matrix is obtained.

6.3.1.6 Getting the Final Score

Finally, all the weightage normalised data values for each row are added to get the overall performance score, and according to the performance scope value, which is more or less, we conclude. For instance, for the best television, the performance scope must be high. The more the value of the score, the better the television.

6.3.2 DATASET

The Heart Disease UCI dataset [33] provided on Kaggle by Ronit has been considered for the MCDM-based heart disease prediction model implementations, and the features and their descriptions are mentioned in Table 6.1. The dataset contains 14 attributes and 303 patient details. In addition, there are eight categorical and six numeric characteristics.

TABLE 6.1

Description of All the Features for the Heart Disease UCI Dataset

S. No.	Feature	Description
1.	Age	Age of the patient
2.	Sex	1 = Male and 0 = female
3.	C_p	Chest pain type
4.	trestbps	Resting blood pressure
5.	Chol	Serum cholesterol
6.	fbs	Fasting blood sugar larger than 120 mg/dL (1 = true and 0 = false)
7.	restecg	Resting electrocardiographic result (1 anomaly)
8.	thalach	Maximum heart rate received
9.	exang	Exercise-induced angina (1 = yes)
10.	oldpeak	ST depression induced by exercise
11.	slope	Slope of peak exercise ST
12.	ca	Number of major vessels
13.	thal	Thalassemia (3 = normal, 6 = fixed defects, 7 = reversible defect)
14.	target	1 or 0

This dataset contains patients whose age lies between 29 and 79 years. The gender value for male patients is 1, and that for female patients is 0. The dataset considered the following four types of chest pains:

Category 1: angina due to stenosis of coronary arteries due to a decreased blood flow through the heart muscle.

Category 2: angina chest pain due to emotional or mental stress.

Category 3: non-angina chest pain, which may be due to different reasons and often may not be caused by actual cardiac disease.

Category 4: asymptomatic chest pain, which may not be a heart attack indication.

The fourth attribute trestbps is the measure of the resting blood pressure. The Chol is the level of cholesterol. The fbs implies fasting blood sugar, which is 1 where blood sugar seems to be under 120 mg/dL and 0 if it is higher. The restecg is the resting electrocardiographic result, and the thalach is the highest heart rate. The exang is the exercise-induced angina, identified as 1 in pain and 0 in painless. The oldpeak is ST depression induced by exercise. The slope is ST segment's slope peak value. The ca is the handful of major fluoroscopically coloured vessels, and the thal is the minutes of the exercise in the test period. The last attribute target is the attribute of class. The class attribute is 1 for patients diagnosed with heart disease and is 0 for normal. Table 6.2 shows which features were used as Beneficial and which were used as non-beneficial.

TABLE 6.2
Beneficial and Non-Beneficial Features Categorisation

S. No.	Feature	Beneficial/Non-Beneficial/Not Required
1	Age	Beneficial
2	Sex	Not required
3	C_p	Non-beneficial
4	trestbps	Beneficial
5	Chol	Beneficial
6	fbs	Beneficial
7	restecg	Beneficial
8	thalach	Non-beneficial
9	exang	Beneficial
10	oldpeak	Beneficial
11	slope	Non-beneficial
12	ca	Beneficial
13	thal	Non-beneficial
14	target	Not required

6.4   RESULTS AND ANALYSIS

This task helps identify potential patients who may face the adverse effects of coronary artery disease during the associated decade. This may help to take precautions and then try to avoid the patient's risk of coronary artery disease. Table 6.3 demonstrates the first ten rows of the original data provided after removing unnecessary columns such as sex, as having heart disease does not depend on that much on sex compared to other factors. We also quantified the data to avoid values of NaN (not a number) and to make the best use of the information.

6.4.1   APPLYING VECTOR NORMALISATION

Table 6.4 shows the vector normalised decision matrix after executing vector normalisation on the data. To make further processing of the data easier, the entire data are transformed into values between 0 and 1.

The weight vector for each feature calculated using the entropy method is as in Table 6.5. The total of all the weightage vectors is a perfect 1, which implies the weightage is divided properly.

Table 6.6 shows the weightage normalised decision matrix after having the data multiplied by the corresponding weightage.

After the calculation of the weightage normalised matrix, the performance score is calculated for each patient and the result is stored into a list containing the patient number and the score, i.e. chances of having a heart disease. The top 10 patients at risk are as in Table 6.7.

TABLE 6.3
First Five Rows of the Original Data After Cleaning

	Age	C_p	trestbps	Chol	fbs	restecg	thalach	exang	oldpeak	slope	ca	thal
0	63	3	145	233	1	0	150	0	2.3	0	0	1
1	37	2	130	20	0	1	187	0	3.5	0	0	2
2	41	1	130	204	0	0	172	0	1.4	2	0	2
3	56	1	120	236	0	1	178	0	0.8	2	0	2
4	57	0	120	354	0	1	163	1	0.6	2	0	2
5	57	0	140	192	0	1	148	0	0.4	1	0	1
6	56	1	140	294	0	0	153	0	1.3	1	0	2
7	44	1	120	263	0	1	173	0	0.0	2	0	3
8	52	2	172	199	1	1	162	0	0.5	2	0	3
9	57	2	150	168	0	1	174	0	1.6	2	0	2

TABLE 6.4

Vector Normalised Decision Matrix

	Age	C_p	trestbps	Chol	fbs	restecg	thalach	exang	oldpeak	slope	ca	thal
0	0.065665	0.878033	0.062734	0.053193	0.850929	1.000000	0.943077	0.000000	0.084860	1.000000	0.0	0.024008
1	0.038565	0.918688	0.056244	0.057074	1.000000	0.922848	0.929036	0.000000	0.129135	1.000000	0.0	0.048015
2	0.042734	0.959344	0.056244	0.046572	1.000000	1.000000	0.934728	0.000000	0.051654	0.924835	0.0	0.048015
3	0.058369	0.959344	0.051918	0.053878	1.000000	0.922848	0.932451	0.000000	0.029516	0.924835	0.0	0.048015
4	0.059411	1.000000	0.051918	0.080816	1.000000	0.922848	0.938148	0.100504	0.022137	0.924835	0.0	0.048015

TABLE 6.5

Weight Vector of Each Feature in the Data

S. No	Feature Name	Weight Vector/Objective Weightage
1	Age	0.163042951736507
2	C_p	0.0213819246810200
3	trestbps	0.163998993720759
4	Chol	0.161655366074361
5	fbs	0.0017594949069709835
6	restecg	0.0230130475795685
7	thalach	0.0413288283367588
8	exang	0.0645213341671867
9	oldpeak	0.0930074857760602
10	slope	0.0365357169050051
11	ca	0.0707082111313622
12	thal	0.159046644984438

TABLE 6.6
Weightage Normalised Decision Matrix

	Age	C_p	trestbps	Chol	fbs	restecg	thalach	exang	oldpeak	slope	ca	thal
0	0.010706	0.018774	0.010288	0.08599	0.001497	0.023013	0.038976	0.000000	0.007893	0.036536	0.0	0.003818
1	0.006288	0.009224	0.009224	0.057074	0.001759	0.021238	0.038396	0.000000	0.012010	0.036536	0.0	0.007637
2	0.006968	0.009224	0.009224	0.046572	0.001759	0.023013	0.038631	0.000000	0.004804	0.033790	0.0	0.007637
3	0.009517	0.008514	0.008514	0.053878	0.001759	0.021238	0.038537	0.000000	0.002745	0.033790	0.0	0.007637
4	0.009687	0.008514	0.008514	0.080816	0.001759	0.021238	0.038772	0.006485	0.002059	0.033790	0.0	0.007637

TABLE 6.7
Top 10 Patients with Their Scores

Patient Number	Score of the Patient
223	0.1941477609813658
204	0.1921294052906392
220	0.1914940437542815153
250	0.19061991947815077
195	0.1855499027582605
246	0.18534659393063301
193	0.1853337781132175
221	0.18494765553720152
217	0.18487812191434894
295	0.1846962581401698

6.4.2 Applying Enhanced Accuracy Normalisation

Table 6.8 shows the result of applying the enhanced accuracy normalisation, i.e. enhanced normalised decision matrix.

The weight vector for each feature was calculated using the entropy method.

Table 6.9 shows the weightage normalised decision matrix after having the data multiplied by the corresponding weightage. After the calculation of the weightage normalised matrix, the performance score is calculated for each patient. The result is stored in a list containing the patient number and the score, i.e. chances of having a heart disease. The top 10 patients at risk of heart disease are as in Table 6.10.

Both the results achieved in Tables 6.7 and 6.10 were compared to see how many patients match in a certain range. When first 250 patients were checked, an accuracy of 83.2% and 80% of the data is compared an accuracy of 82.23140495867769%, which implies that the overall accuracy of the results is around 80% is quite decent.

6.5 CONCLUSION AND FUTURE SCOPE

The risk of developing cardiac problems can be significantly decreased by leading a healthy lifestyle, although diet and genetic predisposition can increase the risk. Obesity, high blood pressure, uncontrolled diabetes, and a diet rich in saturated fats are characteristics of food-related risk factors. The likelihood of living a relatively normal life in the future can be considerably increased by receiving a quick diagnosis of heart disease or prompt treatment after an attack. The problem of heart disease prediction using an MCDM approach is addressed in this research. It's helpful to identify and help patients or citizens having a chance of having a heart disease within them and to take preventive measures against the health issues to keep the person healthy. There can be more features that can affect the chances for one to have a heart disease. Family ancestry of coronary illness can likewise be an explanation for building up a coronary illness as referenced before. Along these lines, this information of the patient can likewise be incorporated for further expanding the precision of the 7 model. The review also shows that the MCDM can be applied to a wide range of areas in health care, using a variety of methodological approaches. The CVD illness expectation can be possible utilising other machine learning and deep learning calculations. Further research is needed to develop clinical practice guidelines for the proper use and reporting of MCDM methods, and the results can be compared and improvised. It can also be concluded that there is immense scope for machine learning in estimating the risks of heart-related conditions. There are many more algorithms that work exceptionally well in some cases, but fail to give accurate results in others. In addition, the experimental results show that the algorithm predicts the probability of cardiac diseases with about 80% accuracy. Additionally, there are a number of potential enhancements in MCDM that might be addressed to increase accuracy.

TABLE 6.8

Enhanced Accurate Normalised Decision Matrix

	Age	C_p	trestbps	Chol	fbs	restecg	thalach	exang	oldpeak	slope	ca	thal
0	0.997959	1.000000	0.997345	0.996562	1.000000	1.01250	1.002182	0.995098	0.997506	1.004717	0.995964	0.990385
1	0.994167	1.003413	0.996621	0.996738	1.022222	1.00235	1.000629	0.995098	0.998273	1.004717	0.995964	0.995192
2	0.994751	1.006826	0.996621	0.996261	1.022222	1.01250	1.001259	0.995098	0.996930	1.000000	0.995964	0.995192
3	0.996938	1.006826	0.996139	0.996593	1.022222	1.00625	1.001007	0.995098	0.996546	1.000000	0.995964	0.995192
4	0.997084	1.010239	0.996139	0.997819	1.022222	1.00625	1.00625	1.000000	0.996419	1.000000	0.995964	0.995192

TABLE 6.9

Weightage Normalised Decision Matrix

	Age	C_p	trestbps	Chol	fbs	restecg	thalach	exang	oldpeak	slope	ca	thal
0	0.064127	0.106631	0.064086	0.064035	0.156847	0.117490	0.087132	0.063951	0.064096	0.084112	0.063998	0.063655
1	0.063884	0.106996	0.064039	0.064046	0.160332	0.116765	0.086997	0.063951	0.064145	0.084112	0.063998	0.063964
2	0.063921	0.107360	0.064039	0.064015	0.160332	0.117490	0.087051	0.063951	0.064059	0.083717	0.063998	0.063964
3	0.064062	0.107360	0.064008	0.064037	0.160332	0.116765	0.087029	0.063951	0.064034	0.083717	0.063998	0.063964
4	0.064071	0.107724	0.064008	0.064115	0.160332	0.116765	0.087084	0.064266	0.064026	0.083717	0.063998	0.063964

TABLE 6.10

Top 10 Patients with Their Scores

Patient Number	Score of the Patient
195	1.0058538964039174
204	1.0057191856440346
193	1.0056566450800815
166	1.0056485342307022
256	1.0056218294994457
191	1.0056214466015871
246	1.0056036393091996
165	1.0055978855247527
174	1.0055794449188837
233	1.005559331112576

REFERENCES

[1] Nalluri, S., Saraswathi, R. V., Ramasubbareddy, S., Govinda, K., & Swetha, E. (2020). Chronic heart disease prediction using data mining techniques. In *Data Engineering and Communication Technology* (pp. 903–912). Springer, Singapore.

[2] Virani, S. S., Alonso, A., Benjamin, E. J., Bittencourt, M. S., Callaway, C. W., Carson, A. P., et al. (2020). Heart disease and stroke statistics—2020 update: A report from the American Heart Association. *Circulation*, 141(9), e139–e596.

[3] U.S. National Library of Medicine. (2021). Heart disease prevention. *MedlinePlus*. Retrieved January 21, 2022, from https://medlineplus.gov/howtopreventheartdisease.html.

[4] World Health Organization. (2004). *The Atlas of Heart Disease and Stroke*.

[5] Nason, E. (2007). *An Overview of Cardiovascular Disease and Research. WR-467-RS*.

[6] R. Chadha, S. Mayank, A. Vardhan, and T. Pradhan. (2016). Application of data mining techniques on heart disease prediction: A survey. In *Emerging Research in Computing, Information, Communication and Applications* (pp. 413–426). Springer India, New Delhi.

[7] A. A. Abdul-Aziz, P. Desikan, D. Prabhakaran, and L. F. Schroeder. (2019). Tackling the burden of cardiovascular diseases in India: The essential diagnostics list. *Circulation: Cardiovascular Quality and Outcomes*, 12(4), e005195.

[8] Salman, O. H., Zaidan, A. A., Zaidan, B. B., Naserkalid, F., & Hashim, M. (2017). Novel methodology for triage and prioritizing using "big data" patients with chronic heart diseases through telemedicine environmental. *International Journal of Information Technology & Decision Making*, 16(05), 1211–1245.

[9] Raju, C., Philipsy, E., Chacko, S., Suresh, L. P., & Rajan, S. D. (2018). A survey on predicting heart disease using data mining techniques. In *2018 Conference on Emerging Devices and Smart Systems (ICEDSS)* (pp. 253–255). IEEE.

[10] Marimuthu, M., Deivarani, S., & Gayathri, R. (2019). Analysis of heart disease prediction using various machine learning techniques. In *Advances in Computerized Analysis in Clinical and Medical Imaging* (pp. 157–168). Chapman and Hall/CRC.

[11] Singh, P., Singh, S., & Pandi-Jain, G. S. (2018). Effective heart disease prediction system using data mining techniques. *International Journal of Nanomedicine*, 13, 121–124. https://doi.org/10.2147/ijn.s124998.

[12] Pathak, A. K., & Valan, J. A. (2020). A predictive model for heart disease diagnosis using fuzzy logic and decision tree. In *Smart Computing Paradigms: New Progresses and Challenges* (pp. 131–140). Springer, Singapore.

[13] Mehmood, A., Iqbal, M., Mehmood, Z., Irtaza, A., Nawaz, M., Nazir, T., & Masood, M. (2021). Prediction of heart disease using deep convolutional neural networks. *Arabian Journal for Science and Engineering*, 46(4), 3409–3422.

[14] Isola, R., Carvalho, R., & Tripathy, A. K. (2012). Knowledge discovery in medical systems using differential diagnosis, LAMSTAR, and k-NN. *IEEE Transactions on Information Technology in Biomedicine*, 16(6), 1287–1295.

[15] Ayon, S. I., Islam, M. M., & Hossain, M. R. (2020). Coronary artery heart disease prediction: a comparative study of computational intelligence techniques. *IETE Journal of Research*, 1–20.

[16] Hassani, M. A., Tao, R., Kamyab, M., & Mohammadi, M. H. (2020). An approach of predicting heart disease using a hybrid neural network and decision tree. In *Proceedings of the 2020 5th International Conference on Big Data and Computing* (pp. 84–89).

[17] Shah, D., Patel, S., & Bharti, S. K. (2020). Heart disease prediction using machine learning techniques. *SN Computer Science*, 1(6), 1–6.

[18] Nagaprasad, S., Pushpalatha Reddy, T., & Naga Lakshmi, S. (2020). Heart disease prediction propagation approach. *International Journal of Machine Learning and Networked Collaborative Engineering*, 4(2), 72–77. Retrieved from https://www.mlnce. net/index.php/Home/article/view/139.

[19] Gavhane, A., Kokkula, G., Pandya, I., & Devadkar, K. (2018). Prediction of heart disease using machine learning. In *2018 Second International Conference on Electronics, Communication and Aerospace Technology (ICECA)* (pp. 1275–1278). IEEE.

[20] Tasnim, F., & Habiba, S. U. (2021). A comparative study on heart disease prediction using data mining techniques and feature selection. In *2021 2nd International Conference on Robotics, Electrical and Signal Processing Techniques (ICREST)* (pp. 338–341). IEEE.

[21] Islam, M. T., Rafa, S. R., & Kibria, M. G. (2020). Early prediction of heart disease using PCA and hybrid genetic algorithm with k-means. In *2020 23rd International Conference on Computer and Information Technology (ICCIT)* (pp. 1–6). IEEE.

[22] Vafaei, N., Ribeiro, R. A., & Camarinha-Matos, L. M. (2018). Data normalisation techniques in decision making: case study with TOPSIS method. *International Journal of Information and Decision Sciences*, 10(1), 19–38.

[23] Ma, W., Luo, X., & Jiang, Y. (2017). Multicriteria decision making with cognitive limitations: A DS/AHP-based approach. *International Journal of Intelligent Systems*, 32(7), 686–721.

[24] Lepri, B., Oliver, N., Letouzé, E., Pentland, A., & Vinck, P. (2018). Fair, transparent, and accountable algorithmic decision-making processes. *Philosophy & Technology*, 31(4), 611–627.

[25] Vafaei, N., Ribeiro, R. A., & Camarinha-Matos, L. M. (2016). Normalization techniques for multi-criteria decision making: analytical hierarchy process case study. In *Doctoral Conference on Computing, Electrical and Industrial Systems* (pp. 261–269). Springer, Cham.

[26] Lakshmi, T. M., & Venkatesan, V. P. (2014). A comparison of various normalization techniques for order performance by similarity to ideal solution (TOPSIS). *International Journal of Computing Algorithm*, 3, 882–888.

[27] Chowdhury, P., & Paul, S. K. (2020). Applications of MCDM methods in research on corporate sustainability: a systematic literature review. *Management of Environmental Quality: An International Journal*, 31, 2, 385–405.

[28] Asadabadi, M. R., Chang, E., & Saberi, M. (2019). Are MCDM methods useful? A critical review of analytic hierarchy process (AHP) and analytic network process (ANP). *Cogent Engineering*, 6(1), 1623153.

[29] G. Adunlin, V. Diaby, and H. Xiao, (2015) Application of multi-criteria decision analysis in health care: a systematic review and bibliometric analysis. *Health Expect*, 18(6), 1894–1905.

[30] T. D. C. Frazão, D. G. G. Camilo, E. L. S. Cabral, and R. P. Souza. (2018). Multi-criteria decision analysis (MCDA) in health care: A systematic review of the main characteristics and methodological steps. *BMC Medical Informatics and Decision Making*, 18(1), 90.

[31] Aouadni, S., Rebai, A., & Turskis, Z. (2017). The meaningful mixed data TOPSIS (TOPSIS-MMD) method and its application in supplier selection. *Studies in Informatics and Control*, 26(3), 353–363.

[32] N. Vafaei, R. A. Ribeiro, and L. M. Camarinha-Matos (2016). Normalization techniques for multi-criteria decision making: Analytical hierarchy process case study. In *Technological Innovation for Cyber-Physical Systems* (pp. 261–269). Cham: Springer International Publishing.

[33] Heart Disease UCI. (2018). *Kaggle*. https://www.kaggle.com/ronitf/heart-disease-uci.

7 State-of-the-Art Literature Review on Classification of Software Reliability Models

Vikas Shinde, S.K. Bharadwaj and D.K. Mishra
Madhav Institute of Technology & Science

CONTENTS

7.1 INTRODUCTION

In the last few decades, almost all organisations have totally been dependent on software systems. The fast growth of the competitive world requires more effective software systems that have more influence on our daily life. Such importance of software insists that developers develop fault-free software. It is expected to design more reliable software and predict their high level of accuracy. The failure of software has given more impact as consequences of enterprises growth in terms of revenue.

DOI: 10.1201/9780367816414-7

The fault rate of the software can be reduced in a systematic way by incorporating various steps such as prevention of errors, fault detection/identification, removal and operational environment. Simultaneously, hardware components are continuously monitored and change from time to time and when required for maintaining the reliability of the system. Nowadays, various software metrics are available to identify the reliability of the software; however, software reliability prediction models are not enough to provide/judge the effective prediction. Software products are used in their high-quality demand with more accuracy and zero tolerance of fault in space science, aviation, defence, high-level data warehouses, etc. In the present scenario, the responsibility of developers is more challenging to provide quality software, which requires to improve software quality, such software quality provides the assurance of software in sequence of reliability. Again, it is a question of common objective how to prophesy and estimate the software reliability. The requirement study of the software reliability with the consideration for improving the quality of software which is challenging for researchers in reliability literature. A general software development process is depicted in Figure 7.1, which describes the user's requirements as well as specifications, design and testing for quality checking. Thereafter, it comes into the operational mode, which explains the predicted rendering and function of the system.

In case the performance of the software deviates from its specification, this indicates that failure of the software occurs; the failure of the software is known as software fault. This fault is called a software failure in the program. Generally, software error is known as software bug also. In general, the life cycle of software considers productivity, quality, cost and delivery. The reliability of the software is totally dependent on the input, conditional logics and coding of the program. The software quality elements are shown in Figure 7.2.

The key concern of software reliability is to describe the quality, measurement and assessment. There are set of questions that arise during the development of software. Consequently, such questions give/provide quality software where failure occurrence has zero tolerance. In Figure 7.3, we demonstrate the process of software assessment/operation.

In Figure 7.4, we describe the software behaviour, which is based on input–programme–output (I–P–O). The depicted mapping considers everything from input data to output data.

7.1.1 Basic Terminology

We highlight few important terminologies that are used throughout the chapter. Such terminology is also discussed by various researchers, including Lyu (1994), Pham (2006), Singpurwalla and Wilson (1999), and Hanagal and Bhalerao (2021).

Software failure describes the incapability of performing the proposed task in a specified framework.

Failure means the condition of the program which does not fulfil the desirable objective/ prescribed requirement.

Bug is a mistake in a program that generates the fault in software under the condition in which the program runs. If such an error cannot be removed immediately, the same fault occurs successively.

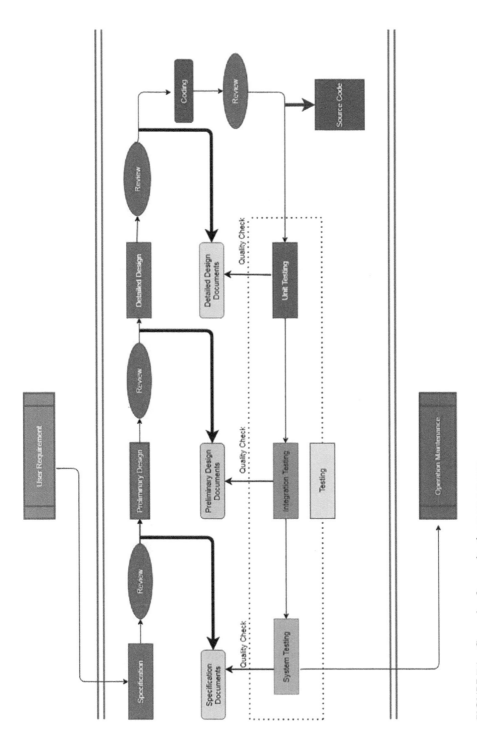

FIGURE 7.1 General software development process.

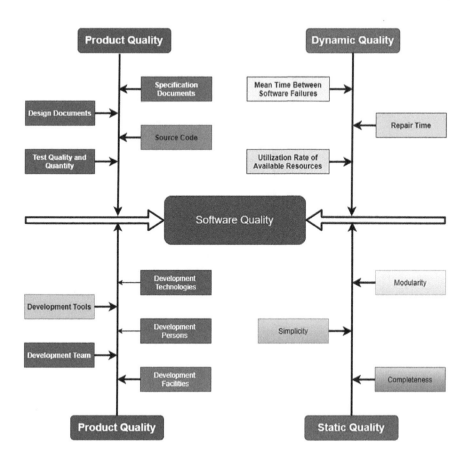

FIGURE 7.2 Software quality.

Debugging is the identification and modification of bug or error in software. There are two kinds of debugging.

Faultless debugging gives an assurance if the software fails, immediately the fault is removed.

Imperfect debugging means there is no guarantee that the error that appeared can be removed from the software. In the process of fault removal, the number of faults may increase or decrease.

Error generation is a testing evolution. Under the removal of original fault, immediately new faults come into existence.

Failure rate describes the period in which failure occurs in a non-uniform manner.

Constant failure rate is the period in which failure occurs in a uniform rate.

Failure density expresses the life of component at any point wherein the number of faults gradually improves with respect to time.

Fault rate function explains the likelihood that a failure unit of time $(t, t+\Delta t)$ occurs in the interval.

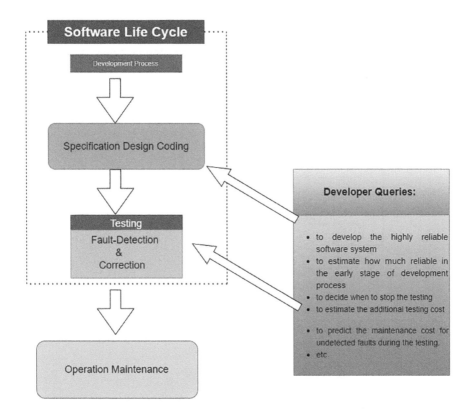

FIGURE 7.3 Software life cycle.

Fault intensity function provides the pace of transformation in the accumulative fault function. This can be evaluated by several faults unit of time because the fault intensity changes over time.

Mean value function represents the average collective failure connected with every point of time.

7.2 SOFTWARE RELIABILITY MODELS

The aim of a software reliability model (SRM) is to estimate real-time problem on a large scale, which assists the management to take appropriate decisions. So, the human life can be squared as well as farad in finance can be controlled and there are endless applications serve by such models. Actually, the testing of software in different phases is very important. Critically, developers have to identify the failure occurrence during the testing. There are so many ways through which software is tested stepwise, such as proper identification of the problem, coding, fault deduction and removal of fault. Altogether, the aim is to develop fault-free or zero tolerance software. Today, many organisations are dependent on software; therefore, software

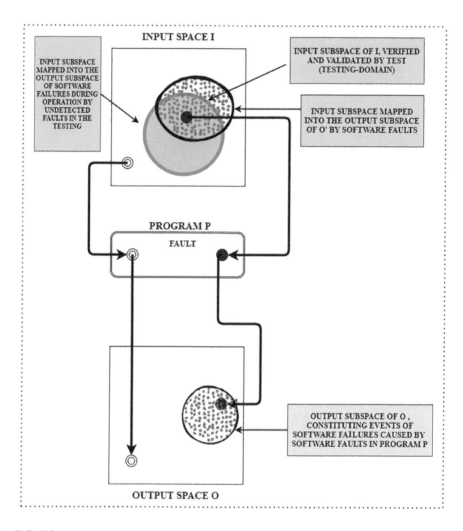

FIGURE 7.4   Input-output program structure.

reliability plays an important role. Actually, SRMs assess the present condition and predict the future condition of software system. Statistically, it is explained in terms of probability of failure and fault forecasting. Quality of any software is working efficiently for a specific period of time after that failure occurs this shows that the particular software has contributed its 100% output but it is not necessary however during this period. During the testing of software, run-time errors occur, which have to be addressed by the developers, and their successive failure/passing rate must be examined for future improvement. SRMs are of service in daily life, for example air traffic control system, space programme, military operations, bank services and human life at very high cost. Generally, SRGMs are classified into two sections: dynamic and static. In Figure 7.5, we focus on the categorisation of these models.

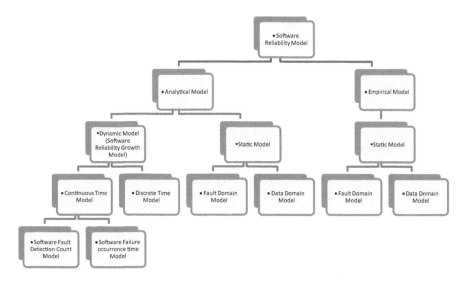

FIGURE 7.5   Types of software reliability models.

7.2.1   SOME MORE APPLICABLE SOFTWARE RELIABILITY MODELS

Some important SRMs are studied in different frameworks, which are discussed below.

7.2.1.1   Non-Homogeneous Poisson Process (NHPP)

Non-homogeneous Poisson process models use analytical approach, which describes the software failure phenomenon during the testing. The focus of NHPP model is to obtain the mean value function to represent the predicted several faults skilled up to a definite time $\{N(t): t \geq 0\}$, where $N(t)$ is the collective number of faults identified in time t. NHPP models are classified into two: finite and infinite. In NHPP models, the predicted number of faults detected given an incalculable of testing time, will be finite failure, while the incalculable fault models assume incalculable faults would be examined in incalculable testing. Various models listed under NHPP are used to evaluate the reliability; some of them are as follows:

- Generalised Goel
- Goel-Okumoto
- Musa-Okumoto
- Modified Duane
- Logistic growth
- Gompertz
- Delayed S-shaped
- Infection S-shaped
- Yamada exponential
- Yamada Raleigh
- Yamada imperfect debugging model 1

- Yamada imperfect debugging model 2
- PNZ model
- Pham Zhang IFD
- Zhang-Teng Pham
- P-Z model.

7.2.1.2 S-Shaped Software Reliability Growth Model

The failure observation phenomenon is described by S-shaped curves and mixed exponential curves. It is considered the defects identified throughout the checking and operating phases are evaluated correctly and eliminated completely. This process is known as perfect debugging. However, debugging process is not always accurate and it depends on developer skill, expertise, data set, real testing and operation environment. Sometimes, testing team is not able to detect and remove the fault; such phenomenon is known as imperfect debugging. There is a chance that while correcting the error/fault another fault/error may occur in the software, this situation is known as error generation and such model is known as error generation model. Herein, all failure content augments checking headways because new faults are inducted into the system while eliminating the actual fault. So, many times the fault removal process is not countered precisely, which creates the imperfect debugging environment. The imperfect debugging should be taken care of perfectly to estimate reliability assessment measures more accurately (Figure 7.6).

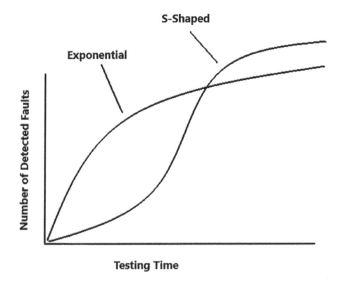

FIGURE 7.6 Graphs of exponent and S-shaped functions.

7.2.1.3   Imperfect Debugging

Indeed, this is the fact that no software can run without having any bug (fault); during the testing of software, it goes through various steps of testing where the process of debugging is performed. There is certain possibility during the debugging operation that a new error will appear while removing an extant fault. More faults may be detected during the execution of software because of imperfect debugging. The situation of perfect debugging is ideal under so many considerations. Testing and process of debugging is a very necessary step in the fruitful enhancement of software systems. Under the process of debugging, some records of execution have been analysed, such as resource consumption, raw data faults and time needed to fix fault. These records create the important information that can help the developers. On this basis, the project manager estimates the improvement of the checking stage, evaluates whether the allocated checking sources are enough, investigates the fault method and determines the optimal time to stop checking and discharge the software. Along with this data collection, failures of software are responsible for providing the quality software. Several researchers used parametric and non-parametric techniques to carry out the effective forecasting of software failure process. In recent years, the most important technique has been the knowledge-dependent system, which can be estimated by any non-linear continuous function dependent on the provided data design. Moreover, the process of debugging is classified into three stages: error detection, fault isolation and fault removal/correction.

Actually, the debugging process includes analysing and extending the given program that does not meet the specifications, in order to develop new programs that satisfy the specifications. This process identifies the accurate nature of the error and thereafter makes it correct. During the development of software, the errors are identified in two ways, namely program proving and program testing. Program proving is based on mathematical logics, and program testing is more realistic and heuristic. However, no one gives complete guarantee to provide an accuracy of the program. The experimental metric of quality is widely used in software testing.

Some authors use Bayesian theory to identify several faults in software. For the removal of faults, Bayesian approach has been utilised. Traditional forecast estimation models offer tools for risk estimation and allow decision-makers to include historical data with subjective estimation.

7.2.1.4   Soft Computing

An important factor, reliability is for obtaining software quality as well as software developers and software users. It concerns fault-free software operation for a certain period of time in a certain environment. In this sequence, the problem is based on two major factors. One is finding a mathematical model to explain the software faults, and the other is assuming the parameters of the model that has depicted the foremost fitting with software fault. Various algorithms are very effective and capable optimisation methods with non-linear, multi-objective, non-differentiable functions. By these algorithms are to obtain the maximum likelihood estimation for the NHPP software reliability, etc., it is the most significant non-functional needs for software reliability. Precisely, evaluating the reliability for amenity-oriented system is impossible.

Software reliability estimation is applicable in many areas such as maintenance and production purposes, failures examination and requirements of manufacturing process in software reliability. It is so difficult/hard to explain the software reliability system dependent on principle-based techniques. The main challenge of such methods is to determine the reliability of complex system by conventional methods. In this way, the authenticity of software reliability computationally highly depends on the software computing as well as prior hypothesis, but this hypothesis may not always be capable in the realistic environment of the systems which go in front of incorrect reliability hypothesis. So the traditional methods are used by adopting soft computing methods. Consequently, soft computing methods permit to assume the reliability by failure behaviour tendencies.

7.3   CLASSIFICATION OF SOFTWARE RELIABILITY MODELS

The classification of SRMs is dependent on software development life cycle (SDLC). This classification of models helps us to select the appropriate category as per the requirement. After studying various models, we shall have to choose a more suitable and realistic model than the existing ones by identifying the unrealistic assumptions made for these existing models; such selection of model gives more exposure to help the management take right decision. On the other hand, the wrong selection of model provides unrealistic and faulty results.

7.3.1   ANALYTICAL MODEL

The analytical modelling of software reliability is developed in two ways: dynamic and static models. The following steps pertain to developing the analytical model.

- Properly define the problem along with conditions.
- Write the analytical model and test procedure.
- Data collection for tuning the parameters.
- Performance analysis.

7.3.2   DYNAMIC OR PROBABILISTIC MODEL

The time-dependent behaviour of the software failure is considered under dynamic model. Probabilistic models express the failure occurrence and fault deduction and removal just because of the involvement of randomness events. Such a model can be divided into different categories as follows:

- Error seeding
- Failure rate
- Curve fitting
- Reliability growth
- Program structure
- Input domain
- Execution path
- Non-homogeneous Poisson process
- Markov
- Unified and Bayesian.

Dynamic models are divided in two categories, namely discrete time model and continuous time model, which are as follows.

7.3.2.1   Discrete Time Models

Discrete time model represent the failure of software in discrete time, and time interval may be definite or arbitrary, so discrete time models are further split in two: definite time interval model and arbitrary time interval model. Here, the list of few models is given.

- Shooman model
- LaPadula model
- Moranda-geometric-Poisson model
- Schneidewind model.

7.3.2.2   Continuous Time Models

Continuous time models describe the failure of software in continuous time. Further, such kind of model is bifurcated in two categories: independently distributed inter-failure time models and independent and identical error behaviour models. Few models under this category are mentioned below:

- Jelinski-Moranda model
- Schick-Wolverton model
- Littlewood-Verrall model
- Wagoner model
- Lipow model
- Moranda-geometric model
- Goel-Okumoto model (NHPP)
- Goel-Okumoto model (imperfect debugging)
- Shantikumar model (Non-homogenous Markov process model).

Here we focus on some frequently used models triggered by the researchers that dominate the leading contribution in increasing the quality of software reliability.

7.3.3 Static or Deterministic Model

These models have time-dependent behaviour in which software development considers the following steps such as observing different sets of error and software failure data including input data.

Generally, the static type of model is applied to study the following:

- Program count by the various operators, operands and instructions.
- Flow chart of a program through which execution of path for branches can be counted.
- The data passing and sharing in a program by flow of data.
- No randomness is involved in deterministic model.
- Performance indices can be determined by analysing the program consistency.

Two models are defined in this category: Halstead's software science model and McCabe's cyclomatic complexity model. Halstead's software science model is applied to obtain several faults in the program; however, McCabe's cyclomatic complexity model is used/applied to obtain upper bound on the number of tests in a program. We list few static models here:

- Mills model
- Lipow model
- Basin model
- Nelson model.

Various kinds of models are developed as per the requirement of organisations/-industries/individual customers; these models are divided step-wise according to which category they fall in fact each model has to follow a phase wise development life cycle of software reliability (Figure 7.7).

7.4 PROCEDURES AND TOOLS

The assessment procedure of software reliability data analysis is depicted in Figure 7.8. Yamada et al. (1989) proposed a software reliability evaluation tool wherein analysis and assessment procedures are mentioned. The flow chart of the process includes the program package using the comfortable language of developer. SRET involves three SRGMs based on NHPP, such as exponential, delayed S-shaped and inflection S-shaped models, along with two deterministic models such as the logistic and Gompertz growth curve models. Soft reliability evaluation tool is helpful and useful for software engineers/software developers to perform the software reliability assessment/evaluation in a systematic and interactive manner without knowing the details of data analysis. The flow chart of the procedures is depicted.

A list of the tools of the SRM is discussed in Table 7.1.

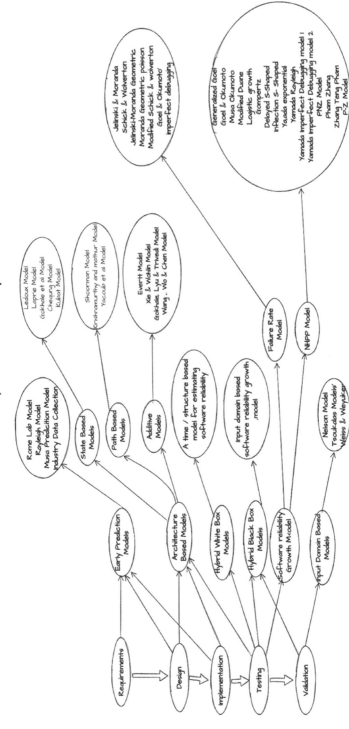

FIGURE 7.7 Life cycle of software reliability development models.

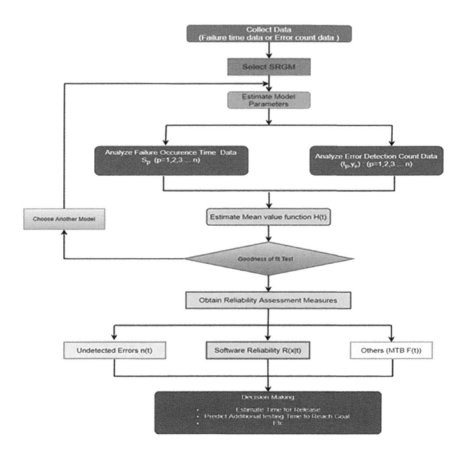

FIGURE 7.8   SRET evaluation process.

TABLE 7.1

Software Reliability Assessment Tools

Tool	Integrated SRGM	Developer	Reference
SORPS	• Delayed S-shaped SRGM • Inflection S-shaped SRGM • Exponential SRGM	IBM	Obha (1984)
SPARC	• Logistic growth curve model • Gompertz growth curve model • Delayed S-shaped SRGM	Toshiba	Nakamura et al. (1985)
Software Reliability Evaluation Program	• Exponential SRGM • Delayed S-shaped SRGM • Logistic growth curve model • Gompertz growth curve model	Toshiba Engineering	Komuro (1987)
SOREM	• Exponential SRGM • Delayed S-shaped SRGM • Logistic growth curve model • Gompertz growth curve model	NEC	Uemura et al. (1990)

7.5 LITERATURE REVIEW

For more than last four decades, researchers have thoroughly analysed and examined various techniques to get best service, which is fault-free or zero error tolerance software. The related literature is analysed by the researchers considering different frameworks such as SRM, software release time, imperfect debugging, application of soft computing in the development of reliability models and availability of software and hardware. Thoroughly, we review so many articles. Schick and Wolverton (1978) introduced an SRM that is divided in two categories: data domain and time domain. The detailed characterisation of advantages and disadvantages of models is also emphasised. Goel and Okumoto's (1979a,b) Markovian model is described with debugging, and all the faults cannot be removed with assurance as and when they are detected. Obha (1984) explained the enhancement of traditional software reliability interpretation models by manufacturing estimates based on more practical. Yamada (1984) described an S-shaped SRM and its application. Defective amending and software availability models are also mentioned. Goel (1985) focused on main modelling methods and gave a vital analysis of the restraints and usability of the models during the software enhancement. Goel (1985) and Hsu and Huang (2011) analysed an SRM for complex system under certain assumptions and limitations. Software reliability is varied on operational environment; proper metric is required to analyse the degree of correctness and quality of software, which definitely enhance the testing efforts. Bittanti et al. (1988) expressed a model of software reliability that is sufficiently feasible to explain a variety of reliability trends. Obha and Chou (1989) explained the enhancement of traditional SRGM by eradication of the arbitrary presumption that faults in a program can be completely doffed. Kapur and Garg (1990) explained a SRGM under defective amending on the basis of non-homogeneous Poisson process, and the parameters of the models are evaluated. Yamada (1991) explained the statistical extent and judgment of software reliability. The methods are dependent on SRGMs introduced in Japan. Kapur and Garg (1992) explained an optimal joint plan explained for such a SRGM depending on the cost-reliability criterion. Van Pul (1992) proposed the utilisation of software reliability theory, which is important for asymptotic conditions of the model. Sahinoglu (1992) suggested the random variable X/sub rem/, which is the residual number of software failures. Yamada et al. (1992, 1993) suggested two software reliability judgment models with defective amending by presuming that new errors are occasionally proposed when the errors initially hidden in a software system are emendated and doffed during the checking stage. Kuo and Yang (1995) examined the prediction of future failure time and future reliability. Wood (1996) depicted that predictions from simple models of fault occurrence times correspond sanely well with the field data to evaluate across multiple software releases to find the suitable models and obtain belief in the results. Pham and Zhang (1997) encapsulated SRMs which are dependent on a non-uniform Poisson method. Chen and Singpurwalla (1997) suggested the unite the various different approaches to growth of reliability models and gives customary design under the software reliability. Gokhale et al. (1998) gave an analytical approach to architecture-based software reliability. The heterogeneous software system based on architecture is used. Pham and Pham (2000) proposed two models prophesy average time when later fiasco depended on Bayesian strategy.

Tokuno and Yamada (2000) provided an SRM to explain the imperfect debugging atmosphere; in this way, the fault correction activity corresponding to failure of each software package is not accomplished completely. Popstajanova and Trivedi (2001) evaluated the software behaviour from the initial stage to the design stage to implementation and final deployment. Classification and identification is proposed for architecture-based models. Chang (2001) investigated an SRGM under non-homogeneous Poisson process to estimate the unknown parameters by least squares method for change point model; such a new approach has more applicability in reliability engineering. Pham and Zhang (2003) introduced a SRM that includes the conceal information, and this information very significant for both software developers and software products. Shyur (2003) utilised the failure data set of different projects to examine software reliability growth model (SRGM) and error analysis with the consideration of defective amending and change point problem to estimate the parameters. The error detection is challenging because it is dependent on testing environment, resource allocation and strategy. The classical maximum likelihood method is used. Zhang et al.'s (2003) defective amending is speculated in the view that new errors can be acquainted into the software during amending and the discovered errors may not be doffed totally. Huang et al. (2003) discussed various SRGMs based on NHPP, which are extensively expressed by using arithmetic, geometric and harmonic mean with this general transformation is formulated. Gokhale et al. (2004) determined analytical results for architecture-based SRM forecasting and its performance evaluation. Kapur et al. (2004) SRGM are carried out for distributed development environment. NHPP models consider the software system that includes a finite number of reused subsystems, which has adverse impacts on the system; however, the new subsystem provides the growth uniformly. Jeske and Zhang's (2005) SRGM with various frameworks is examined by using architecture-based SRM in different test environments along with practical problems wherein the diversified behaviour of test and operational profiles is discussed. Huang (2005) proposed new theorems and data collection for software testing in real-time applications, and logistic testing effort function and change point parameter are applied. Actually, fault detection is a change between the processes of software development. Teng and Pham (2006) produced a new technique for estimating software reliability in the meadow surrounding, which gives a workable means to model consumer environments and moreover generates alterations to the reliability prophecy for alike software goods. Zhang and Pham's (2006) methodology of field failure rate prediction is explained, and the test data and filed data are explored, which has more concern of SRGMs. Particularly, the mismatch of operational profile of the test and filed environments is discussed. Predicting field failure rates include that fault removals in the field are usually non-instantaneous and fixes of certain faults reported in the field can be delayed. Wang et al. (2006) explained the architecture-based approach for modelling the software reliability, and the different characteristics of architectural styles are used to incorporate the non-uniform behaviour of software embodying heterogeneous architecture. Singh et al. (2007) retrospect in what way distinct SRGMs have been grown where error recognition method is based not only on the several remaining error content but also on the time of trials, and observe in what way these models can be explained as the postponed error recognition model by applying a prolong aftermath aspect. Su and Huang

(2007) introduced an artificial neural network-dependent method for software reliability evaluation and modelling. Pham's (2007) imperfect debugging of the software should be identified, and the parameters on which it depends are examined/tuned. Fault detection is critical exercise during the development of software. Kapur et al. (2008) evaluated the fault detection of the software at the time of release. Ramasamy and Gopal (2008) proposed Goel-Okumoto SRGMs to examine the failure intensity function by using shifted Weibull function. Yang (2010) studied data-driven SRMs with multiple-delayed-input single-output architecture with the consideration of recent failure. Hsu and Huang's (2011) adaptive approach of path testing is used for modular software system, which indeed is helpful and useful to estimate the studied software reliability. Hsu and Huang (2010) suggested a modified genetic algorithm to obtain the parameters of SRGMs. Trials dependent on real software fiasco data are accomplished, and outcomes depict that the propounded genetic algorithm is highly efficient and quicker than conventional genetic algorithms. Huang and Lyu (2011) proposed a powerful technique to use under testing and operational phases for software reliability assessment and forecasting. NHPP-based SRGMs are derived using unified theory with the idea of multiple change points is also demonstrated. Rahamneh et al.'s (2011) genetic programming is applied to obtain the best performance of SRGM in an automated way. The proposed model is compared with Yamada S-shaped model and few NHPP models, which validate that the obtained results are superior. Ahmad et al. (2011) presented a software enhancement of expense curve and to compare the efficacy for the suggested model and another extant model. Mahapatra and Roy (2012) deployed the modified J-M model which explained the flawed debugging method with usability of the model has been depicted on the failure data set of Musa. Subburaj et al. (2012) described NHPP and SRGM to analyse the failure data adequately for improving the quality of debugging such as imperfect debugging, perfect debugging and efficient debugging. Lai and Garg (2012) studied extant SRMs dependent on NHPP, which allege to enhance software quality by efficient recognition of software faults. Okamura et al. (2013) suggested SRGMs, which are mathematically manageable and have enough capability of appropriate to the software failure data with the given parameter estimation algorithm for the SRGM with normal distribution. Peng et al. (2014) developed a testing method for analysing the imperfect debugging with the consideration of detection and correction. Kaur and Sharma (2015) discussed the comparative study between failures and accuracy estimation. Li et al. (2015) established the idea to incorporate the S-shaped function into non-homogenous Poisson process software reliability model for imperfect software debugging. Wang et al. (2015) studied log logistic distribution for evaluating the imperfect software debugging. Kim et al. (2015) propounded an efficient method to obtain the parameters of SRGM applying a real-valued genetic algorithm (RGA). Present SRGMs crave the appraisement of the parameters like as total number of unsuccessful or the unsuccess detection rate applying numerical methods or least square estimation. Jin and Jin (2016) examined the enhancement and utilisation of a swarm intelligent optimisation algorithm, specifically/as a quantum particle swarm optimisation (QPSO) algorithm, to improve these parameters of SRGMTEF as well as comparative relation with other existing models. Li and Yi (2016) introduced an improved SRGM to reconsider the reliability of open-source software (OSS) systems to certify the model's portrayal applying various real-world data. Hanagal and Bhalerao (2016a, b) discussed the concepts of SRGMs, which have been a moderately fruitful tool in technology. Wang et al. (2016) discussed the novel idea to enhance the optimised SRM wherein function implement successively with exponential distribution to best fit a logarithmic deviation between observed value and estimated value from fault data set. Optimised models fit the fault data set accurately in a better way than traditional models based in software testing. Chatterjee and Shukla (2016) considered two kinds of software faults, such as independent and dependent. Also, the fault reduction rate is treated as a proportionality function. The performance of the model is much better on failure data set, which is evaluated on the bases of predicted and estimated number of faults. Li and Pham (2017) considered fault detection based on testing coverage under the uncertainty of operating condition. Li and Pham (2017) suggested a new model with the deliberation of the faulty recognition rate dependent on the inspecting coverage and pondered on cover ID subject to the ambiguity of operating environments. Erto et al.'s (2018) generalised inflection S-shaped SRGM is discussed with its properties. Maximum likelihood estimators are used to formulate the model parameters. Optimal release time of the software is also emphasised. Hanagal and Bhalerao (2018) considered an S-shaped SRGM with the concept of error generation based on NHPP. Estimate whether the data is performed using a maximum likelihood technique. Choudhary et al. (2018) propounded an efficient parameter appraisement method for SRGMs applying firefly algorithm. Software unsuccessful rate with respect to time has been a leading apprehension in the software industry. Li and Pham (2019) applied to elicit models that include the ambiguity of operating environments, which gives the pliability in considering a distinct faulty recognition rate and random environmental element and so on. Hanagal and Bhalerao (2019) suggested a model for comparison with standard models on the basis of different data sets. Kaliraj et al. (2020) exposed SRMs in different frameworks for their utility and applicability. Tahere and Yaghoobi (2020) propounded a modified differential evolution (MDE) algorithm for resolving a exalt amplitude non-linear optimisation task. The topic obtained maximum likelihood estimation (MLE) for the parameters of a NHPP software reliability model. Amar et al. (2021) described the hybrid reliability-based design optimisation (RBDO) techniques used in HEMT method to enhance its performance and reliability are described. The use of RBDO methods needs the enhancement and coupling of two models. Lin and Chen (2021) propounded two new models including time-varying unsuccessful intensity in each stage. These models receive the plan from the accelerated failure-time models. And modification component is brought in to develop the relationship between two consecutive position parameters. Nor et al. (2021) suggested the direction of the similarities between the included domains and the difficulties in reliability science carrying out are unveiled. The techniques deployed in respective industry are described, with each stamina and frailty investigated, together with useful examples. Shorthill et al. (2021) suggested a new and collective reach to the reliability analysis case study of BAHAMAS, which is depicted to be a pliable tool whose usability is constructed to handily include traditional probability hazard evaluation. Robinson et al. (2021) discussed a software construction design, which gives a viable system for building self-steady models and coercing feedback to limit analyst fault. The data of various SRMs are expressed by a pie chart in Figure 7.9.

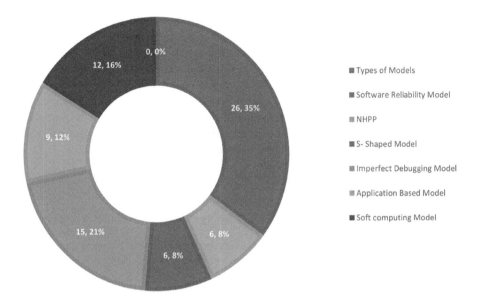

Types of Models

Software Reliability Model

NHPP

S- Shaped Model

Imperfect Debugging Model

Application Based Model

Soft computing Model

FIGURE 7.9 Classification of software reliability models.

7.6 CONCLUSIONS

Software reliability models have been developed in the last four decades. The development of these models is under different assumptions, conditions and environment. It is generally observed that these models may be in conflict with one another/each other. There should be an effective and efficient procedure to recognise/identify the most suitable/appropriate model for specific kind of problem along with the condition under which it performs. The performance of the software is a big concern because it immediately reflects the quality; such quality assures the customer satisfaction and further revenue generation for company. There is a need to develop such technique which supports evaluating many errors that remain in the application of software at the release time. At that point if the reliability is below the acceptable level, then it refers to re-testing until the desired level of reliability is achieved definitely increase the test expenditure amount for debugging process. Kubat and Kochc (1983) proposed at the release time various test assay manners to identify many errors in the software. Singpurwalla (1991) investigated a model to predict the reliability of a software program, which really helps to determine the attempt needed before stopping, checking and amending of the software. Acton et al. (2014) developers of the software products are keenly interested to provide the best quality of software by defining the problem and evaluate metrics concern to quality and again use such metrics to enhance the quality of software products. Various matrices and models were proposed to measure the quality of the software. During the development phase of software the fault detection can be emphasised so that the failure occurrence has been removed. Software reliability is a very important measure in planning as well as controlling the sources through the development; therefore, high quality of software can be developed. The best quality of the software is required by the society because it is widely used in finance, health, aviation, defence and many more areas.

In sequencing to achieve the best reliability of the system, a broad test plan is essential, which ensures us that all the necessary components have been incorporated and tested. The main issues has to comprise how to quantify reliability, how to design test, cost and resource constraints, what are the inference of test failures and which type of the test should be re-run so by adopting the following correction the fault may be removed or rectified. These all concern issues of uncertainties in the quality of software and testing efforts have been focused by Yamada (1993).

REFERENCES

Acton, D., Kourie, D. G., Watson, B. W. (2014): Quality in software development: A pragmatic approach using metrics, *SACJ*, 52, 1–12.

Aggarwal, A. G. (2019): Multi release reliability growth modeling for open source software under imperfect debugging, In *System Performance and Management Analytics*, Springer, Singapore, pp. 77–86.

Ahmad, N., Khan, M. G. M., and Rafi, L. S. (2011): Analysis of an inflection S shaped software reliability model considering log-logistic testing-effort and imperfect debugging. *International Journal of Computer Science and Network Security*, 11(11), 161–171.

Amar, A. Radi, B., and Hami, A. E. (2021): Reliability based design optimization applied to the high electron mobility transistor (HEMT), *Microelectronics and Reliability*, 124, 114299.

Bittanti, S., Bolzern, P., Pedrotti, E., Pozzi, N., and Scattolini, R. (1988): A flexible modeling approach for software reliability growth. In G. Goos, & J. Harmanis (Eds.), *Software Reliability Modelling and Identification*, Springer Verlag, Berlin, pp. 101–140.

Chang, Y. P. (2001): Estimation of parameters for nonhomogeneous Poisson process software reliability with change-point model, *Communications in Statistics – Simulation and Computation*, 30(3), 623–635.

Chatterjee, S., and Shukla, A. (2016): Change point-based software model under imperfect debugging with revised concept of fault dependency, *Journal of Risk and Reliability*, 230(6), 579–597.

Chen, Y., and Singpurwalla, N. D. (1997): Unification of software reliability models by self-exciting point processes, *Advances Application Probability*, 29, 337–352.

Choudhary, A., Baghel, A. S., Sangwan, O. P. (2018): Parameter estimation of software reliability model using firefly optimization. In: *Data Engineering and Intelligent Computing*, Springer, Singapore, pp. 407–415.

Erto, P., Giorgio, M., and Lepore, A. (2018): The generalized inflection S-shaped software reliability growth model. *IEEE Transaction in Reliability*, 69(1), 228–244.

Goel, A.L (1985): Software reliability models: Assumptions, limitations and applicability. *IEEE Transactions on Software Engineering*, 12(2), 1411–1423. https://doi.org/10.1109/tse.1985.232177.

Goel, A. L., and Okumoto, K. (1979a): A Markovian model for reliability and other performances measures of software system. In: *Proceedings of COMPCON*, IEEE Computer Society Press, Los Angeles, CA, pp.769–774.

Goel, A. L., and Okumoto, K. (1979b): Time dependent error-detection rate model for software reliability and other performance measures. *IEEE Transactions on Reliability*, 28(3), 206–211.

Gokhale, S., Hong, W.E., Trivedi, and K., Horgan, (1998): An analytical approach to architecture based software reliability predication, *Proceeding of the 3rd IEEE International Computer Performance and Dependability Symposium*, 13–22.

Gokhale, S. S., Wong, W.E., Horgan, J. R., and Trivedi, K. S. (2004): An analytical approach to architecture-based software performance and reliability prediction. *Performance Evaluation*, 58(4), 391–412, https://doi.org/10.1016/j.peva.2004.04.003.

Goos, G. and J. Harmanis (Eds.), *Software Reliability Modeling and Identification*, Springer, Berlin, pp. 101–140.

Hanagal, D. D., and Bhalerao, N. N. (2016a): Analysis of NHPP software reliability growth model. *International Journal of Statistics and Reliability Engineering*, 3(2), 53–67.

Hanagal, D. D., and Bhalerao, N. N. (2016b): Modeling and statistical inference on generalized inverse Weibull software reliability model. *Journal of Indian Society of Probability and Statistics*, 17(2), 145–160.

Hanagal, D. D., and Bhalerao, N. N. (2017): Modeling on extended inverse Weibull software reliability growth model. *Indian Association for Product, Quality and Reliability Transactions*, 42(2), 75–95.

Hanagal, D. D., and Bhalerao, N. N. (2018): Analysis of delayed S-shaped software reliability growth model with time dependent fault content rate function. *Journal of Data Science*, 17(3), 573–589.

Hanagal, D. D., and Bhalerao, N. N. (2019): Modeling on generalized extended inverse Weibull software reliability growth model, *Journal of Data Science*, 16(4), 857–878.

Hanagal, D. D., and Bhalerao, N. N. (2021): *Software Reliability Growth Models*, Infosys Science Foundation Series Published by Springer.

Hsu, C.J. and Huang, C.Y. (2010): A study on the applicability of modified genetic algorithms for the parameter estimation of software reliability modeling. In: *Computer Software and Applications Conference (COMPSAC), IEEE 34th Annual, IEEE*, pp. 531–540.

Hsu, C.J., and Huang, C.Y. (2011): An adaptive reliability analysis using path testing for complex component-based software systems. *IEEE Transactions on Reliability*, 60(1), 158–170.

Huang, C.Y. (2005): Performance analysis of software reliability growth models with testing-effort and change-point. *Journal of Systems and Software*, 76(2), 181–194.

Huang, C.Y., and Lyu, M.R. (2011): Estimation and analysis of some generalized multiple change-point software reliability models. *IEEE Transactions on Reliability*, 60(2), 498–514.

Huang, C. Y., Lyu, M. R and Kuo, S. Y. (2003): A unified scheme of some non homogenous Poisson process models for software reliability estimation. *IEEE Transactions on Software Engineering*, 29(3), 261–269.

Jeske, D. R., and Zhang, X. (2005): Some successful approaches to software reliability modeling in industry. *Journal of Systems and Software*, 74(1), 85–99.

Jin, C. and Jin S.W. (2016): Parameter optimization of software reliability growth model with S-shaped testing-effort function using improved swarm intelligent optimization, *Applied Soft Computing*, 40, 283–291.

Kaliraj, S., Vivek, D., Kannan, M., Karthick, K., Dhasny Lydia, M. (2020): Critical review on software reliability models: Importance and application of reliability analysis in software development. *Materials Today: Proceedings*.

Kapur, P. K., and Garg, R. B. (1990): Optimal software release policies for software reliability growth model under imperfect debugging. *Recherche Opertionnelle/Operations Research (RAIRO)*, 24, 295–305.

Kapur, P. K., and Garg, R. B. (1992): A software reliability growth model for an error removal phenomenon, *Journal of Software Engineering*, 7, 291–294.

Kapur, P. K., Garg, R. B., and Kumar, S. (1999): *Contributions to Hardware and Software Reliability*, World Scientific Publishing Co. Ltd, Singapore.

Kapur, P.K., Goswami, D.N., and Gupta, A. (2004): A software reliability growth model with testing effort dependent learning function for distributed systems. *International Journal of Reliability, Quality and Safety Engineering*, 11(4), 365–377.

Kapur, P.K., Gupta, D., Gupta, A., Jha, P.C. (2008): Effect of introduction of fault and imperfect debugging on release time, *Ratio Mathematica*, 18, 62–90.

Kaur, D., & Sharma, M. (2015): Software reliability models: Time between failures and accuracy estimation, *International Journal of Computer Science and Information Technologies*, 6(4), 3370–3373.

Kim, T., Lee, K. and Baik, J. (2015): An effective approach to estimating the parameters of software reliability growth models using a real-valued genetic algorithm. *Journal of Systems and Software*, 102, 134–144.

Komuro, Y. (1987): Evaluation of software products' testing phase application to software reliability growth models. In *Reliability Theory and Application*, ed. Osaki, S. and Cao, J. World Scientific, Singapore, pp. 188–194.

Kubat, P. and H. S. Kochc (1983): Pragmatic testing protocols to measure software reliability. *IEEE Transactions on Reliability*, R-32(4), 338–341.

Kuo, L., and Yang, T. Y. (1995): Bayesian computation of software reliability. *Journal Computations Graphical Statistics*, 4, 65–82.

Lai, R., and Garg, M. (2012): A detailed study of software reliability models. *Journal of Software*, 7(6), 1296–1306.

Li, F., and Yi, Z. L. (2016): A new software reliability growth model: Multi generation faults and a power-law testing-effort function. *Mathematical Problems in Engineering*, 1–13.

Li, Q. Y., and Pham, H. (2017): NHPP software reliability model considering the uncertainty of operating environments with imperfect debugging and testing coverage. *Applied Mathematical Modelling*, 51, 68–85.

Li, Q. Y., and Pham, H. (2019): A generalized software reliability growth model with consideration of the uncertainty of operating environments. *IEEE Access*, 7, 84253–84267.

Li, Q.Y., Li, H.F., and Lu, M.Y. (2015): Incorporating S-shaped testing-effort functions into NHPP software reliability model with imperfect debugging, *Journal of Systems Engineering and Electronics*, 26(1), 190–207.

Lin, K., and Chen Y. (2021): Two new multi-phase reliability growth models from the perspective of time between failures and their applications, *Chinese Journal of Aeronautics*, 34, 341–349.

Lyu, M. R. (1996): *Handbook of Software Reliability Engineering*, McGraw-Hill, New York.

Lyu, M. R. (1996): *Handbook of Software Reliability Engineering*, IEEE Computer Society Press, CA, 1996.

Mahapatra, G. S., and Roy, P. (2012): Modified Jelinski-Moranda software reliability model with imperfect debugging phenomenon. *International Journal of Computer Applications*, 48, 38–46.

Musa, J. D., Lannino, A., and Okumoto, K. (1987): *Software Reliability: Measurement, Prediction and Application*, McGraw-Hill, New York.

Nakamura, H., Uchihira, N., Satoh, M. and Mizutani, H. (1985): A method of predicting software quality prior to the testing phase, *Nikkei Electronics*, 25(2), 233–260.

Nor, A.K.M., Pedapati, S.R., Muhammad, M. (2021): Reliability engineering applications in electronic, software, nuclear and aerospace industries: A 20 year review (2000–2020), *Ain Shams Engineering Journal*, 12(3), 3009–3019.

Obha, M. (1984): Software reliability analysis models. *IBM Journal of Research Development*, 28(4), 428–443.

Obha, M., and Chou, X. (1989): Does imperfect debugging affect software reliability growth models? In *Proceedings of the 4th International Conference on Software Engineering, Pittsburg*, pp. 237–244.

Okamura, H., Dohi, T., Osaki, S. (2013): Software reliability growth models with normal failure time distributions. *Reliability Engineering & System Safety*, 116, 135–141.

Peng, Y.F., Li, W.J., Zhang, Q.P., Hu, (2014): Testing effort dependent software reliability model for imperfect debugging process considering both detection and correction. *Reliability Engineering & System Safety*, 126, 37–43.

Pham, H. (2000): *Software Reliability*, Springer, New York.

Pham, H. (2006): *System Software Reliability*, Springer, New York.

Pham, H. (2007): An imperfect debugging fault-detection dependent parameter software. *International Journal of Automation and Computing*, 4(4), 325–328.

Pham, H., and Zhang, X. (1997): An NHPP software reliability model and its comparison. *International Journal of Reliability, Quality and Safety Engineering*, 4, 269–282.

Pham, H., and Zhang, X. (2003): NHPP software reliability and cost models with testing coverage, *European Journal of Operational Research*, 145(2), 443–454.

Pham, L., and Pham, H. (2000): Software reliability models with time-dependent hazard function based on Bayesian approach. *IEEE Transactions on Systems, Man and Cybernetics, Part A: Systems and Humans*, 30(1), 25–35.

Popstajanova, K., and Trivedi, K. (2001): Architecture based approach to reliability assessment of software systems. *Performance Evaluation*, 45(2), 179–204.

Rahamneh, Z.A.L., Reyalat, M., Sheta, A. F., Bani-Ahmad, S., Al-Oqcili, S. (2011): A new software reliability growth model: Genetic-programming-based approach. *Journal of Software Engineering and Applications*, 4(8), 476–481.

Ramasamy, S. and Gopal, G. (2008): A software reliability growth model addressing learning. *Journal of Applied Statistics*, 35, 1151–1168.

Robinson, A.C., Drake, R. R., Swan, M. S., Bennett, N. L., Smith, T. M., Hooper, R., Laity, G. R. (2021): A software environment for effective reliability management for pulsed power design. *Reliability Engineering & System Safety*, 211, 107580.

Sahinoglu, M. (1992): Compound-Poisson software reliability model. *IEEE Transactions on Software Engineering*, 18(7), 624–630.

Schick, G. J., and Wolverton, R. W. (1978): An analysis of competing software reliability models. *IEEE Transactions on Software Engineering*, 4(2), 104–120.

Shorthill, T., Bao, H., Zhang, H., Ban, H. (2021): A novel approach for software reliability analysis of digital instrumentation and control systems in nuclear power plants. *Annals of Nuclear Energy*, 158, 108260.

Shyur, H.J. (2003): A stochastic software reliability model with imperfect debugging and change-point. *Journal of Systems and Software*, 66(2), 135–141.

Singh, V. B., Yadav, K., Kapur, R., and Yadavalli, V. S. S. (2007): Considering the fault dependency concept with debugging time lag in software reliability growth modeling using a power function of testing time. *International Journal of Automation and Computing*, 4(4), 359–368.

Singpurwalla, N. D. (1991): Determining an optimal time interval for testing and debugging software. *IEEE Transactions on Software Engineering*, SE-17(4), 313–319.

Stark, G., and Lyu, M. (1996): Software reliability tools. In *The Handbook of Software Reliability Engineering*, McGraw-Hill, New York.

Su, Y. S. and Huang, C. Y. (2007): Neural-network-based approaches for software reliability estimation using dynamic weighted combinational models. *Journal of Systems and Software*, 80(4), 606–615.

Subburaj, R, Gopal, G. and Kapur, P. K. (2012): A software reliability growth model for estimating debugging and the learning indices. *International Journal of Performability Engineering*, 8(5), 539–549.

Tahere, and Yaghoobi (2020): Parameter optimization of software reliability models using improved differential evolution algorithm. *Mathematics and Computers in Simulation*, 177, 46–62.

Teng, X., & Pham, H. (2006): A new methodology for predicting software reliability in the random field environments. *IEEE Transactions on Reliability*, 53(3), 458–468.

Tokuno, K., and Yamada, S. (2000): An imperfect debugging model with two types of hazard rates for software reliability assessment and assessment. *Mathematical and Computer Modelling*, 3, 343–352.

Uemura, M., Yamada, S. and Fujino, K. (1990): Software reliability evaluation method: Application of delayed S-shaped NHPP model and other related models. *Proceedings of the Annual Reliability and Maintainability Symposium*, 467–472.

Van Pul, M. C. (1992): Asymptotic properties of a class of statistical models in software reliability. *Scandinavian Journal of Statistics*, 19(3), 235–253.

Wang, J., Wu, Z., Shu, Y., and Zhang, Z. (2015): An imperfect software debugging model considering log-logistic distribution fault content function. *Journal of Systems and Software*, 100, 167–181.

Wang, J., Wu, Z., Shu, Y. and Zhang, Z. (2016): An optimized method for software reliability model based on non homogeneous Poisson process. *Applied Mathematical Modelling*, 40(13–14), 6324–6339.

Wang, W. L., Pan, D. and Chen, M. H. (2006): Architecture-based software reliability modeling. *Journal of Systems and Software*, 79(1), 132–146.

Wood, A. (1996): Predicting software reliability. *IEEE Computer*, 11, 69–77.

Xie, M. (1991): *Software Reliability Modelling*, World Scientific, Singapore.

Xie, M. (2000): Software reliability models- past, present and future. In N. Limnios & M. Nikulin (Eds.), *Recent Advances in Reliability: Methodology, Practice and Inference*. Springer Science Business Media, LLC, New York.

Yamada, S. (1989): *Software Reliability Assessment Technology*, HBJ, Japan, Tokyo.

Yamada, S. (1991): Software quality/reliability measurement and assessment: Software reliability growth models and data analysis. *Journal of Information Processing*, 14, 254–266.

Yamada, S. (2014): *Software Reliability Modeling Fundamentals and Applications*. Springer: New York.

Yamada, S., Isozokai, R, and Osaki, S. (1989): A software reliability evaluation tool, SRET. *Transactions on IEICE*, J72-D-I(1), 24–32.

Yamada, S., Obha, M., and Osaki, S. (1984): S-shaped software reliability growth models and their applications. *IEEE Transactions on Reliability*. 33, 289–292.

Yamada, S., Tokuno, K., and Osaki, S. (1992): Imperfect debugging models with fault introduction rate for software reliability assessment. *International Journal of Systems Sciences*, 23(12), 2241–2252.

Yamada, S., Tokuno, K., and Osaki, S. (1993): Software reliability measurement in imperfect debugging environment and its application. *Reliability Engineering & System Safety Journal*, 40(2), 139–147.

Yang, B., Li, X., Xie, M., and Tan, F. (2010): A generic data-driven software reliability model with model mining technique. *Reliability Engineering and System Safety*, 95(6), 671–678.

Zhang, X., and Pham, H. (2000): Comparisons of non-homogeneous Poisson process software reliability models and its applications. *International Journal of Systems Science*, 31(9), 1115–1123.

Zhang, X., and Pham, H. (2006): Software field failure rate prediction before software deployment. *Journal of Systems and Software*, 79(3), 291–300.

Zhang, X., Teng, H. and Pham, H. (2003): Considering fault removal efficiency in software reliability assessment. *IEEE Transactions on System, Management and Cybernetics-Part A*, 33(1), 114–120.

8 Survey on Software Reliability Modelling and Quality Improvement Techniques

Manish Bhardwaj
KIET Group of Institutions, Delhi-NCR

Korhan Cengiz
University of Fujairah

Vineet Sharma
KIET Group of Institutions, Delhi-NCR

CONTENTS

DOI: 10.1201/9780367816414-8

8.1   INTRODUCTION

Desktop computers and the software that runs on them have a tremendous impact in the society [1]. Digital instrumentation has replaced analogue and mechanical components in electronic devices such as autos, washers, TV, gasoline pumps and microwave ovens. The software industry and its related enterprises are growing at a breakneck speed.

Processors and development tools systems are prepared to offer minimalistic design, adaptable handling, a richness of capabilities and a competitive price as the cost of processing continues to decline and the amount of influence rises. Computers and smart materials are fast displacing their mechanical counterparts from the marketplace, just as machinery supplanted handcraft during the Industrial Revolution. Almost all software systems have a high level of dependability as their fundamental dynamic attribute. End-users bear the burden of increased costs as a result of unreliable software [2]. In computing, reliability is a measure of how well customers believe a software application or software system performs in providing the services they require.

Programming is becoming increasingly important in the design of complex frameworks nowadays, as seen by the increasing importance placed on it. Because the product is a scholarly item, it is not constrained by the requirements of the real world, as it would be in a comparable equipment framework, and this is the primary reason for this distinction [3]. Because programming is always performed in the context of a larger framework, the reliability requirements for the framework are sent down to the product component(s) and become the ideal programming reliability requirements.

In terms of programming dependability, it is one of the most fundamental barriers between high-quality programming and high-reliability frameworks. It is defined as the possibility of disappointment-free programming activity occurring in a set climate for a predefined period of time under certain conditions [4]. A product disappointment occurs when the programming's execution deviates from its specifications. It is the result of a product deficiency, also known as a plan deformity, which is performed by a specific contribution to the code during its execution, resulting in a plan deformity.

Programming dependability testing is performed at various points during the process of designing programming for a framework in an effort to determine whether or not the product's steadfast quality requirements have been (or can be) met [5]. The findings of the inquiry serve as input to the architects and as a measure of the overall quality of the programming. The evaluation and expectation activities are the two exercises that are associated with the programming dependability examination. At some point in the course of action, quantifiable anticipating techniques and steadfast quality models are applied to disappointment information gathered from testing or during activity to assess the dependability of software development projects in one way or another [6]. In any situation, evaluation is typically undertaken to determine the level of dependability that has been achieved from a previous point in time to the current point in time. The expectation action, on the other hand, sets the dependability models that will be used for assessment and then uses the available information to predict future reliability. Program dependability models can be assigned in two ways: as dark box models or as white box models, as a general rule. The fundamental distinction between the two is that white box modelling considers the item's framework when determining supportability, whereas black box models do not [7]. The dependability of an assessment or programming architecture is proportional to how well clients believe it provides the forms of assistance that they require. This chapter seeks to provide a study of coding dependability, which has been divided into three sections: demonstrating, estimating or measurement, and upgrades.

Programming is defined as the prospect of long-term programming activity in a preset environment that is devoid of disappointment; reliability of the programming is described in [2]. While electronic and mechanical hardware can age and degrade with time and use, the software that runs on it will remain the same. After a period of time, programming will not change unless it is altered or redesigned with the intention to do so. When it comes to programming quality, reliability is one of the most important characteristics to consider. Other important characteristics to consider include utility, convenience, execution time, workability, capacity, installability, viability and documentation [8].

8.2 RELIABILITY CURVE

Program or application dependability is part of the amount of disappointments faced by an individual consumer of that program or application. When a product is being implemented, it's easy to become disappointed. Programming failure occurs when a customer or client requests or expects a service lives.

A malfunctioning ATM machine, which frequently occurs when the machine does not remember your last withdrawal, will likely make you happy. However, in planes, heart pacemakers and radiation treatment machines, a product error can literally save people's lives by saving them from certain death or serious injury [9]. Figure 8.1 depicts the disappointed characteristics caused by long-term usage of the equipment, also known as the bath that isn't provided by the program. Programming errors have saved a great many people's bend. Section consumption, indeterminate life, and end-of-life or wear out are represented by A, B and C, respectively.

The natural factors that make equipment to wear out do not have any impact on programming unwavering quality. A superior bend is displayed in Figure 8.2 when programming unwavering quality is projected on similar tomahawks. As per dependability assessment focus, 1996, there are two essential varieties between equipment and programming bends. One change is that in the last segment, programming doesn't have a developing disappointment rate as equipment does. Currently, software is becoming dated, and no new ideas for upgrades or modifications exist. That's

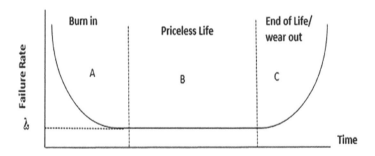

FIGURE 8.1   Hardware reliability in the form of bathtub curve.

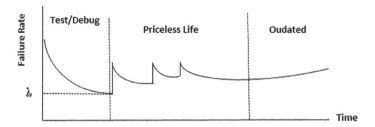

FIGURE 8.2   Curve for software reliability.

why disappointment won't take the place of it. The second contrast is that at the critical life stage, software may see an unusual increase in the rate of disappointment each time an improvement is achieved. It's normal for the disappointment rate to go down after a while due to any defects that were discovered and repaired during the process of improvement.

8.3   REVIEW OF SOFTWARE RELIABILITY MODEL

Various programming dependability models have arisen as individuals attempt to comprehend the characteristics of how and why programming falls flat, and try to evaluate programming dependability. Something like 200 models have been proposed by programming specialists, yet how to gauge programming dependability still remains inexplicable.

In the product improvement measure, it is extremely average to wind up with an item that has numerous configuration absconds, for example deficiencies, or prevalently known as bugs [10]. For a specific contribution to the product, these shortcomings are enacted, bringing about a deviation of the product conduct from its predefined conduct, for example a disappointment. Once disappointments are recognised through the testing interaction and the comparing fault(s) are found, then, at that point expecting that these issues are consummately fixed, for example the way towards fixing a shortcoming, didn't present another shortcoming, programming unwavering quality increments. In the event that the disappointment information is recorded by the same token as far as number of disappointments noticed per given time frame or as far as the time between disappointments, measurable models can be utilised to recognise the pattern in the recorded information, mirroring the development in unwavering quality. Such models are known as programming unwavering quality development models (SRGMs) or development models overall [11]. They are utilised to both anticipate and gauge programming unwavering quality.

All SRGMs are of the discovery type since they just think about disappointment information, or measurements that are accumulated if testing information is not accessible. Black box models don't think about the interior design of the product in unwavering quality assessment and are called as such in light of the fact that they consider programming as a solid element, a black box [12].

In the ensuing bits of this segment, five SRGMs are introduced. These are to be specific the Jelinski-Moranda de-eutrophication model, Nelson model, the Musa essential execution time model, the upgraded NHPP (ENHPP) model and the Littlewood-Verrall Bayesian model [13].

8.3.1   MODEL OF J-M DE-EUTROPHICATION

There will be N programming issues at the beginning of testing, each one distinct from the others and causing disappointment during testing, according to this model. No new allegations are levelled throughout the troubleshooting stage because a previously discovered problem is fixed with confidence in a short amount of time. The rate of product disappointments or hazard work is stated as a percentage of total time between the 1st and ith disappointments (t_i).

$$Z(t_i) = \emptyset [N - (I - 1)].$$

A proportionality $\emptyset$ stable state is where there is a constant value. Remember that this risky work is constant between disappointments, but it lessens in steps after the erasure of each flaw.

8.3.2   MODEL OF ENHANCED NHPP

NHPP (ENHPP) is a system for limiting disappointment that has been enhanced. A special case of the ENHPP model is an NHPP model that has restricted mean esteem capacity. The model's in-depth details explicitly links time-varying test inclusion and blemish defect location.

This model's test inclusion is defined as the ratio of the number of potential defect localities honed by a test to the total number of potential shortcoming destinations [14]. There are "the programme substances addressing either primary or utilitarian programme components whose sharpening is deemed important towards putting up the functional trustworthiness of the product item" as potential defect destinations.

The model makes the accompanying suspicions:

1. "Shortcomings are consistently appropriated over all potential flaw locales".

2. "The likelihood of identifying a shortcoming when an issue site is sharpened at time t is cd(t) = K, (a steady), the shortcoming discovery inclusion".
3. "Deficiencies are fixed consummately".

8.3.3 MODEL OF MUSA EXECUTION TIME

In this model, the assumptions are the same as in the J-M model; however, the cycle shown is the number of disappointments during predefined execution time periods instead of the J-M model. It receives funding for its high-risk work from

$$Z(r) = \emptyset f(N - nc).$$

In this equation, r represents how long it's been since the program was last ran, F represents how often it's run, which indicates a generally constant state, and nc represents how many times it's been modified while it was running $(0, r)$. This model demonstrates how much the risk work is dependent on the execution time.

8.3.4 MODEL OF NELSON

The dependability of the programming is estimated, according to Nelson [15], by running the product for a test with n inputs and seeing how it performs. A random selection of n inputs is made from the information space set $E = (E_i: i = 1,..., N)$, where each E_i represents the arrangement of information esteems intended to be used to complete a run [16]. If ne is the number of information sources that resulted in execution failures, then an unbiased gauge of programming unwavering quality may be calculated from that number

$$\text{RI is } \left\{1 - (ne \, / \, n)\right\} \text{ and so on.}$$

There isn't a single life-sized replica that may be used throughout the entire festival. No model is complete; one life-sized model may be ideal for a certain programming set-up, but it may also be completely off-screen for a variety of other concerns, depending on the situation. The Markovian model is the foundation for the majority of present insightful approaches to dealing with gain dependability measures for application programs, and they are predicated on the chance of superb disappointment time conveyance as their primary premise [17]. The Markovian models are reliant on the fundamental problem of an insurmountably large state space as their starting point.

Although strategies have been provided to improve the dependability of life-sized models of segments that can't be represented by utilising the conventional insightful methods, they are also confronting the state space blast difficulty. An engaging substitute to an insightful model, however, is a recreation life-sized model or model that depicts a technique being described in expressions of its curios, schedule, interrelationships and cooperative efforts in such a methodology that one may simply perform probes on the model, rather than on the actual framework, preferably with an undefined result.

8.3.5 MODEL OF LITTLEWOOD-VERRALL BAYESIAN

All of the previous models assume that information on disappointment is readily available. They, too, use tried-and-true quantifiable approaches such as the maximum likelihood estimation (MLE), in which model bounds, no matter how hazy, are defined and graded based on the available data. The drawback of this approach is that when disappointment information is unavailable, model bounds cannot be examined.

However, MLE techniques are unreliable if there is a lack of available information, as this could lead to faulty or inaccurate assessments. The Bayesian SRGM analyses constant quality development in terms of both the number of issues that have been identified and the absence of disappointment [18]. Bayesian models also assume that the model borders had a previous appropriation without disappointment information, reflecting judgement on obscure historical knowledge, such as a former form and perhaps a well-qualified assessment of the product Bayesian models.

8.3.6 MODEL OF WHITE BOX SOFTWARE RELIABILITY

White box programming dependability models think about the inward design of the product in the unwavering quality assessment rather than discovery models, which just model the associations of programming with the framework inside which it works. The dispute is that discovery models are lacking to be applied to programming frameworks with regard to segment-based programming, expanding reuse of segments and complex communications between these parts in a huge programming framework. Besides, defenders of white box models advocate that dependability models that consider part reliabilities, in the calculation of by and large programming dependability, would give more practical gauges.

The inspiration to create the supposed "engineering"-based models incorporates advancement of methods to dissect performability of programming worked from reused and business off-the rack (COTS) parts, performing affectability examinations, for example contemplating the variety of use unwavering quality with variety in part and interface dependability, and for the ID of basic parts and interfaces [19].

In these white box models, parts and modules are recognised, with the suspicion that modules are, or can be, planned, carried out and tried autonomously. The engineering of the programming is then distinguished, not in the feeling of the customary computer programming engineering, but instead in the feeling of cooperations between parts. The cooperations are characterised as control moves, basically suggesting that the engineering is a control stream chart where the hubs of the diagram address modules and its advances address move of control between the modules [20]. The disappointment conduct for these modules (and the related interfaces) is then indicated as disappointment rates or reliabilities (which are thought to be known or are processed independently from SRGMs). The disappointment conduct is then joined with the engineering to appraise generally programming dependability as an element of segment reliabilities. The manner by which the disappointment conduct is joined with the engineering recommends that three conventional classes of white box programming unwavering quality models exist: way-based models, state-based models and added substance models.

8.4 METRICS OF SOFTWARE RELIABILITY

Estimation is indicated and careful in other designing region and it isn't determined in programming. Albeit bothering, the chase of evaluating programming dependability has not ever stopped. Except if presently, we actually don't have any magnificent method for estimating application dependability. Estimating programming dependability stays a troublesome concern since we don't have a decent method to comprehend the idea of programming. There isn't any clear definition to what aspects are including program unwavering quality. We can't find a suitable answer for measure programming unwavering quality, and the majority of the highlights including programming unwavering quality. It is enticing to quantify whatever disturbing unwavering quality to mirror the highlights, on the off chance that we can't measure unwavering quality right away [21]. The present practices of programming unwavering quality estimation can likewise be separated into four classes [13].

8.4.1 PRODUCT METRICS

Many variables are mentioned in terms of programming complexity, effort to advance and consistency. A programming project's source code is measured in "lines of code", which is also referred to as "kilo lines of code" (KLOC). It's probable, in any event, that there isn't a standard tally mechanism at the present time. Commentary and other non-executable explanations are typically excluded from source code calculations. Programs written in a different language than the one being analysed cannot be accurately analysed using this methodology. Lines of code for Java and C sharp programming will be distinct from one other. This direct approach to code development and maintenance is also being questioned in light of ongoing breakthroughs in code reuse and code cycle technique.

This statistic is used to measure how far a suggested programming effort has progressed based on information sources, yields, ace archival requests, and interfaces reviewed in detail. It's used to gauge how far along a piece of suggested programming is in execution. Once the product's capabilities are known, the technique may be used to estimate the size of the framework. To put it another way, it measures how difficult the application is to use. It is unaffected by the programming language used and makes an informed guess about the presentation that was dropped at the client. There are many corporate applications that use it; however, it is not designed to operate in a logic or real-time setting. Because dependability in programming is closely linked to complexity, the first step in dealing with complexity is to recognise it. To select the amount of product control structure complexity, complexity-oriented measurements use a graphical representation of the code. When it comes to delegating authority, McCabe's intricacy metric is a good one to use.

Test inclusion measurements are a method of assessing deficiency what's more, dependability with the guide of performing tests on program items, in light of the supposition that programming dependability is an element of the segment of application that has been successfully confirmed or set up.

8.4.2   PROJECT MANAGEMENT METRICS

Analysts have understood that on the money organisation can bring about a superior item. Examination has affirmed that a relationship exists between the improvement strategy and the potential to finish items on schedule and inside the liked acceptable targets. The cost increases when designers utilise deficient strategies. Higher dependability may likewise be realised through using higher progress approach, hazard organisation strategy, design organisation approach, etc.

8.4.3   PROCESS METRICS

Focused with the understanding that the nature of the item is a momentary presentation of the measure, measure measurements can be used to appraise, uncover and improve the dependability and top calibre of programming. In International Organization for Standardization ISO-9000 certification is the accepted standard for a family of norms produced by ISO (ISO).

8.4.4   METRICS OF FAULT AND FAILURE

Determining whether the product is getting close to giving a disappointment-free experience will need gathering deficiency and dissatisfaction measures. Customers' screw-ups (or other issues) after delivery are tallied up and assessed in a small method to attain this purpose, as are the number of inadequacies detected during the course of checking out (i.e. before conveyance). As with fault measures, the test technique is very similar in that it can complete all assessments yet lead to disappointment if the looking at condition does not match the total program utility. Buyer feedback on mistakes that occurred after the application's release is most commonly used to calculate customer dissatisfaction levels. Using the disappointment data gathered, it is possible to compute disappointment thickness, mean time between failures (MTBF), or, on the other hand, screw-ups or various boundaries to evaluate or anticipate program unshakable quality. In order to select the most appropriate metric, it is necessary to consider the type of system to which it will be applied, as well as the requirements of the machine area. It is possible that specific dependability measurements for one-of-a-kind sub-programs may be required for specific projects and will be appropriate in some cases. Table 8.1 contains a list of some of the essential measurements that were used to determine the dependability of a program or a programming environment.

In certain situations, framework clients are normally informed about how the framework will fall short in the long run, most likely due to the fact that restarting the framework would be a significant expenditure. During these instances, a measurement based on the cost of disappointment event (ROCOF) or the inference time to disappointment should be employed.

Because there may be a cost associated with failing to provide assistance in some cases, it is imperative that an organisation's framework consistently meets a request on a number of occasions. The number of disappointments experienced over a period of time is less relevant. The probability of disappointment on request (POFOD) measurement must be applied in these situations. It is possible that clients or framework

TABLE 8.1
Software Reliability Metrics

Metrics	Content	Example
MTTF	This is a proportion of the time between noticed framework disappointments. For instance, a MTTF of 500 methods that one disappointment can be anticipated each 500 time units. In the event that the framework isn't being transformed, it is the proportional of the ROCOF.	Frameworks with long exchanges such as CAD frameworks. The MTTF should be more prominent than the exchange time.
ROCOF	This is a proportion of the recurrence of event with which unforeseen conduct is probably going to happen. For model, a ROCOF of 2/100 implies that two disappointments are prone to happen in every 100 functional time units. This measurement is at some point called disappointment power.	Exchange handling frameworks, operating system.
POFOD	This is a proportion of the probability that the framework will bomb when a help demand is made. For instance, a POFOD of 0.0001 implies that one out of 1000 help solicitations might bring about mistake.	Security basic and constant frameworks, for example equipment control frameworks.
AVAIL accessibility	This is a proportion of how possible the framework is to be accessible for use. For instance, an accessibility of 0.998 implies that in each 1000 time units, the framework is probably going to be accessible for 998 of these.	Ceaselessly running frameworks, for example phone exchanging frameworks.

administrators will be reminded on a regular basis that the cycle is imminent when a request for administration is submitted. If the methodology is unavailable, they will bring about some tragedy for themselves. Accessibility (AVAIL) considers the time it takes to re-establish or restart a service.

8.5 IMPROVEMENT TECHNIQUES OF SOFTWARE RELIABILITY

The improvement of programming frameworks includes a grouping of routines for creating new things, which provides numerous alternatives for infusing human fallibilities. Bumbles may first appear during the actual implementation of the approach, when the objectives may be misunderstood or incompletely considered. Because people can't participate in and preserve a communication with perfection, a quality assurance project is used in conjunction with programming enhancement. Spot on developing strategies, such as programming testing or looking at, programming approval and programming check, can help to improve the dependability of programming in general.

8.5.1   SOFTWARE TESTING

Examining or testing a program produced with the assistance of programmers is an exciting stage in the investigation of the program under consideration. During the course of the previous stage of use designing endeavours, the professional makes an attempt to construct programming from a theoretical concept to an evident finished product. To "wreck" the application, the expert assembles a series of tests that are all run at the same time. App development's only destructive stage is testing, or rather the test planning stage of developing applications. Programmers are often optimistic individuals, regardless of their mode of operation.

Testing implies that the engineer discards assumptions about the "correctness" of the programming that has just been written and avoids a conflict of interest that might arise when mistakes are discovered throughout the process. If the examination is conducted in a practical manner, it will be possible to identify a flaw in the application. Giving it a shot demonstrates that product administrations give the impression of being trustworthy when dealing with specific customers, and that social and proficiency criteria appear to have been followed.

8.5.1.1   Principles of Software Testing

Before employing strategies to construct viable programming tests, a programmer must first understand the following fundamental concepts that underlie programming testing:

1. There should be a clear distinction between the tests and the client's requirements in all cases: Code testing's goal is to find and fix mistakes, as we've seen. As a result, the most critical flaws are those that cause the program to fall short of meeting the needs of the customer.
2. Tests should be planned far in advance of the start of testing: When the requirements model is completed, the test planning phase can begin. When the configuration model has been established, the tests can begin in their nitty-gritty detail. All tests can be planned and organised in this manner prior to any code being written in the first place.
3. When it comes to software testing, the Pareto rule is applicable: 80% of all defects found during testing will most likely be visible to 20% of all software segments, according to the Pareto rule. Of course, the real test is to separate these said parts and put them to the test.
4. Testing should begin "in the little" and advance to testing "in the large" as the following: The first tests that are planned and completed are mostly focused on single parts of information. Test centres alter as testing progresses, with the goal of identifying problems in groups of segments and finally in the entire framework being discovered.
5. It is impossible to imagine an exhaustive testing procedure: Even a very well-estimated program will have a large number of way changes, which is particularly significant. As a result, it is difficult to test each of the possible combinations of approaches. You can cover program logic sufficiently and make sure that all conditions are worked out in part level plan before the program begins despite this.

6. To be effective, attempting should be guided by a free outsider: Tests that are performed to their highest potential for uncovering errors, which is the primary goal of testing, are considered to be at their highest potential. The product engineer who created the framework isn't the most qualified person to oversee all of the product's quality assurance tests.

8.5.1.2 Reliability Testing Importance

The device of PC programming has spread into a wide range of extraordinary industries, with its application forming a critical component of mechanical, commercial and military frameworks, among others. Because of its multiple capabilities in defending head programs, programming unshakable quality is currently a study topic of interest in the field of head programming. Regardless of the manner that application design is fitting the fastest construction innovation of the previous century, there isn't a full, rational, quantitative measure that can be used to evaluate their effectiveness. Programming unwavering quality testing is being used as a tool to assist in the examination of these application design advancements.

A thorough evaluation of unshakable quality is essential in order to increase the productivity of application items and the programming progress approach over time. The importance of evaluating programming dependability can't be overstated due to the fact that it is quite useful for application directors and specialists.

Testing is used to ensure that the product's uncompromising quality is maintained:

1. In order to establish an economically sound estimate of how long the application will run without disappointing, a sufficient number of test conditions must be completed for a suitable period of time. We need long-term experiments in order to detect abandons that require some effort in the reasoning process to develop.
2. The dissemination of test occurrences should match the arranged functional profile of the program. The more likely a capacity or subset of the application is executed, the better the level of sweep cases that should be assigned to that capacity or subset.

8.5.2 Type of Reliability Testing

Programming unwavering quality testing incorporates highlight testing, load testing and relapse testing.

Highlight test: Feature testing surveys the highlights outfitted through the product and is completed in the following advances:

- Each activity inside the application is performed once.
- Transaction between the two activities is diminished.
- And every single activity is checked for its appropriate execution.

Burden testing: This study is done in order to determine the success of the program when it is subjected to the greatest amount of responsibility. Any software performs better up to a certain point in terms of measure of responsibility, after which the program's reaction time begins to degrade significantly.

Consider the following example: A website online can be certified to look at the amount of concurrent clients it can possibly support without compromising its effectiveness. This testing is usually beneficial for database administrators and application workers, among other things. Burden examining also necessitates application execution testing, which determines how well a particular program executes when placed under a lot of obligation.

Regression testing: Regression testing is used to determine whether or not any new defects were introduced as a result of previous nasty software fixes. After each trade or replacement in the program's components, a relapse attempt is carried out. These shots are given on an irregular basis based on the size and components of the program in question.

8.5.3   VERIFICATION AND VALIDATION OF SOFTWARE

In the context of application execution, confirmation refers to the arrangement of events that ensures that an application properly executes a specific activity. Approval relates to an additional schedule arrangement that ensures that the application that has been produced is observable in relation to the client's expectations. This is how Boehm expresses it in more detail: "Would we claim that we are constructing the object that is appropriate?" "Would we claim that we are building the correct item?" the group asks.

8.5.3.1   Validation Testing

Once integration testing is complete, the programming has been completely collected together as a group, any interface flaws have been identified and corrected, and a final grouping of utilisation examinations – approval testing – may be initiated.

Approval is successful when the application incorporates features in its design that are reasonable to expect from the client in question. One factor on which an experienced program engineer may disagree is: Who is the mediator of moderate assumptions with absolute certainty?

In the application requirements, simple assumptions are demonstrated by way of example.

8.5.3.1.1   Specification

A document that depicts all of the distinct and distinguishable characteristics of the programming.

8.5.3.2   Criteria of Validation Testing

Programming approval is done through a progression of dark field watches that show congruity with necessities. A test plan traces the classes of checks to be directed, and a test interaction characterises one of a kind test examples to be utilised to represent congruity with necessities. Both the arrangement and the measure are intended to guarantee that each utilitarian prerequisite is fulfilled, all effectiveness necessities are executed, documentation is appropriate and human engineered, and also, different determinations are met (e.g. movability, similarity, mistake recuperation and viability).

After each approval try case has been done, presumably the most two doable specifications exist:

1. The proficiency qualities adjust to determination and are acknowledged, or
2. A deviation from determination is revealed furthermore, and an inadequacy record is made.

8.6 CONCLUSIONS

As for utilising SRGMs for unwavering quality assessment, thought of the model suppositions is significant before a SRGM is applied to disappointment information to guarantee consistency between the model suspicions and relating information. For instance, if a Weibull or a Gamma dissemination fits the recorded disappointment times well, forecasts acquired from a model that accepts a comparative disappointment time conveyance are bound to be nearer to genuine qualities than an expectation from a model that accepts a dramatic disappointment time dispersion.

Further, citing a perception made by Brocklehurst and Littlewood,

> There is no all around worthy model that can be trusted to give exact outcomes in all conditions; clients ought generally doubt claims actually. More awful, we can't distinguish deduced for a specific information source the model of models, assuming any, that will give exact outcomes; we just don't comprehend which components impact model exactness

It is the normal situation that a gathering of development models having comparative suspicions differ in their expectations for similar arrangement of disappointment information and it is likewise the case that every one of the models makes a similar wrong expectation. In such a situation, the expectations from the models are disputable and may just be best utilised for current dependability assessment as opposed to for forecast.

With respect to white box models, most models make the suspicion that segment reliabilities are accessible and disregard the issue of how they can be resolved. This is as yet an open research issue. With shortage of disappointment information in segments, it isn't generally conceivable to utilise SRGMs to assess part reliabilities, for example in Gokhale et al.'s state-based model.

In addition, the presumption of autonomy between disappointments in segments can be disregarded during unit testing, which infers that at this point an unwavering quality development model cannot be utilised to decide segment reliabilities. Between-segment reliance is thought to be non-existent in engineering-based models, which doesn't appear to be an extremely sensible presumption.

The issue emerges when an interface causes blunder engendering between two segments and causes disappointments in the two segments. This negates the presumption of freedom in segment what's more, interface disappointments, and the models are presently not material. The value of engineering-based models, particularly of state-based models, is principally that the system for unwavering quality forecast can likewise be utilised for execution investigation, just as for affectability examinations and in the ID of basic segments.

At last, most models depend on the presence of disappointment information except for Bayesian development models that accept an earlier conveyance for the SRGM boundaries. In any case, these models experience the ill-effects of their unimportance if programming unwavering quality is an element of the reliabilities of its segments and interfaces. This seems, by all accounts, to be the situation with the expanding utilisation of COTS in building programming. The forecast of unwavering quality at the testing stage considers little criticism to the plan measure since testing is excessively far down the computer programming cycle.

In my view, a binding together system that uses programming measurements ahead of schedule during the product designing cycle, disappointment information, when accessible, measure measurements and interaction history to iteratively appraise or anticipate unwavering quality would be of worth in the feeling of early approval of dependability prerequisites, for plan trade-offs and for assessing programming designs. Further, no structure exists, yet that delivers a sensible expectation of programming dependability when information is careless and refines the forecast when information opens up. These are regions that legitimize further examination.

REFERENCES

[1] Aasia Q., Mehraj_Ud-Din D., Quadri S.M.K. (2010), Improving software reliability using software engineering approach – a review, *International Journal of Computer Applications (0975–8887)* 10(5).

[2] ANSI/IEEE. (1991), *Standard Glossary of Software Engineering Terminology, STD-729-1991*, ANSI/IEEE.

[3] Ian S. (2007), *Software Engineering*, Eighth Edition (ISBN 13: 978-0-321-31379-9, ISBN 10: 0-321-31379-8). Pearson Education Limited. China Machine Press.

[4] Jelinski Z. and Moranda P. (1972), Software reliability research, In *Statistical Computer Performance Evaluation*, W. Freiberger. New York: Academic, pp. 465–484

[5] Pan J. (1999), *Software Reliability*, Carnegie Mellon University.

[6] Musa J. D. (1993), Operational profiles in software- reliability engineering, *IEEE Software* 10(2), 14–32.

[7] Musa J. D. *Software Reliability Engineering: More Reliable Software, Faster and Cheaper*, McGraw-Hill. ISBN 0-07-060319-7.

[8] Musa J. D. (1997), Introduction to software reliability engineering and testing, *8th International Symposium on Software Reliability Engineering (Case Studies). November 2–5, 1997*. Albuquerque, New Mexico.

[9] Kumar M., Ahmad N., Quadri S.M.K. (2005), Software reliability growth models and data analysis with a Pareto test effort, *RAU Journal of Research* 5(1–2), 124–128.

[10] Shanmugam, L. and Florence, L. (2012), An overview of software reliability models, *International Journal of Advanced Research in Computer Science and Software Engineering* 2(10), 36–42.

[11] Musa J. D. (1971), A theory of software reliability and its application, *IEEE Transaction on Software Engineering* SE-1, 312–327.

[12] Reliability Analysis Center. (1996), *Introduction to Software Reliability: A State of the Art Review*. Reliability Analysis Center (RAC). http://rome.iitri.com/RAC/.

[13] Raghvendra K. (2013), Approach for enhancing the reliability of software, *Computing, Information Systems, Development Informatics & Allied Research* 4(4).

[14] *Software Reliability Testing*, Retrieved from http://en.wikipedia.org/w/index. php?title=Software_reliability_testing&oldid=645297567, 24th March, 2015.

[15] Nelson E. (1978), Estimating software reliability from test data, *Microelectronics Reliability* 17, 67–74.

[16] Sinha S., Neeraj Kumar G. and Rajib M. (2019), Survey of combined hardware–software reliability prediction approaches from architectural and system failure viewpoint, *International Journal of System Assurance Engineering and Management, Springer; The Society for Reliability, Engineering Quality and Operations Management (SREQOM), India, and Division of Operation and Maintenance, Lulea University of Technology, Sweden* 10(4), 453–474.

[17] Kim T., Lee K. and Baik J. (2015), An effective approach to estimating the parameters of software reliability growth models using a real-valued genetic algorithm, *Journal of Systems and Software* 102, 134–144.

[18] Rana R., Staron M., Berger C., Hansson J., Nilsson M., Törner F., et al. (2014), Selecting software reliability growth models and improving their predictive accuracy using historical projects data, *Journal of Systems and Software* 98, 59–78.

[19] Ullah N., Morisio M. and Vetro A. (2014), A method for selecting software reliability growth model to predict OSS residual defects, *Computer* 1–1.

[20] Park J. and Baik J. (2015), Improving software reliability prediction through multi-criteria based dynamic model selection and combination, *Journal of Systems and Software* 101, 236–244.

[21] Amin A., Grunske L., and Colman A. (2013), An approach to software reliability prediction based on time series modeling, *Journal of Systems and Software* 86, 1923–1932.

9 Multi-Criteria Decision Making for Software Vulnerabilities Analysis

Aarti M. Karande
Sardar Patel Institute of Technology

Padmaja Joshi
CDAC

CONTENTS

9.1 INTRODUCTION

E-platform generates mountains of information. This large information requires attention for security. People and technology both play an equally important role in information security (9). Software vulnerability is a serious problem. Software vulnerability is defined as a flaw within a software system that could cause violation of its own security policies (2). It works as loopholes to steal sensitive data from the

DOI: 10.1201/9780367816414-9

system (3). Vulnerabilities can give access to attackers. They can control the system and execute illegal actions (1). Vulnerability response depends on time, roles, impact on production process, functionalities and operations. Qualitative measurement approach for handling risks gives a brief of the severity, while quantitative approach gives it a score to quantify the severity (24).

9.2   CAUSES OF VULNERABILITIES

Software error is a reproducible defect. Vulnerability density is measured as the number of vulnerabilities per unit size of code (11). Quantitative characterisation requires the use of models to measure the density. This model captures the repeatable behaviour causing the defect and its frequency which are measurable. Major causes of software vulnerabilities are identified as follows:

- **Insecure interaction between components**. This is a problem due to improper neutralisation of special elements used in any software application. Improper alignments or the miscommunication protocol can be a cause for this. Understanding sequential and proper flow among the components may help to solve this problem.
- **Risky analysis**. Improper access to the input memory or functionality, excessive buffer size or data type overflow or wraparound creates the inaccessible or full accessible data. Security should be considered for different data types in the risk analysis. Data type along with data flow among all the modules are needed to be understood and analysed properly to solve this problem.
- **Missing/broken access**. User restrictions are important. Missing authentication for critical function, missing encryption of sensitive data and incorrect permission assignment for critical resource are some of the major examples. This enables untrustworthy agents to perform replay, injection and privilege escalation attacks. Missing communication protocol with standard communication policy will help to manage the complete activity flow with roles assigned will help to solve this problem.
- **Known vulnerable component**. Application components running on open libraries can be exploited by an untrustworthy agent. This can cause access to the serious data or server. Version change or the use of open libraries may directly or indirectly affect the flow of the applications. The use of valid and authenticated libraries will help to solve this problem.
- **Cross-site scripting**. These flaws execute unauthorised scripts on different pages. It may lead to cross-site scripting. This gives access to illegal data. Data across different pages may create invalid accessibility. The use of authenticated scripting with validation will help to solve this problem.
- **Design vulnerabilities**. Development needs to be done as per the requirement specification. If security requirements are not properly handled as per specification, it may impact the execution. When threats are not properly identified at design level, the impact may affect the application module. Architectural usage with the understanding of design of data will help to solve this problem.

- **Implementation vulnerabilities**. If the implementation deviates from the design to solve technical discrepancies, errors are generated besides known errors. If the specified standards for design do not match, then it will impact the implementation of the module. Infrastructure dependencies are needed to be identified and implemented to solve this problem.
- **Operational vulnerabilities**. Software and hardware interactions works are of the operational environment. Physical operating environment can generate vulnerabilities. Even though there are no major issues with overall working of the system, surrounding environmental accessibility may create errors. Proper security constraints may be of help to solve this problem.

9.3 VULNERABILITY DETECTION METHODS

Software vulnerability is both traceable and non-traceable error. The first step in handling error is to detect it before it creates any problems. Detection methods help us identify tools and techniques suitable for analysing functionality, weakness and strength (2). There are two ways for detection. The first is static detection method. It will check type inference, data flow analysis and constraint analysis. The static method works for compila tion phase. The second is dynamic detection method. This method handles running status and monitoring interface. It checks the program's weaknesses without changing the source code (18). Qualitative detection method measures quality parameter affecting the system. It considers the impact of vulner-ability on the system seriously. Quantitative detection method measures the accuracy of impact on the module as per its level of impact.

9.3.1 LIST OF SOFTWARE VULNERABILITY METHODS

Different detection methods used for detecting systems gaps are listed below (2).

- **Fuzzing**. It is a security detection method. This quantifies the impact of vul-nera bility. Fuzzing requires a standard data generation and target monitoring system. It validates the output as per expectation, based on the invalid or ran-dom in put. It performs pertinent test. It focuses only on the executable codes. Life cycle of fuzzing includes first identifying the target output and inputs. They generate fuzzed data, execute fuzzed data, monitor for exceptions and then determine exploitability w.r.t. the target. There are two types of fuzzing.
  - **Black box fuzzing**. Here, the output is evaluated for the target output. Data are randomly generated by modifying the correct data. It fails to understand the actual requirement of the application. New inputs are generated to meet structural specification without any prior knowledge of the program. It modifies well-formed inputs and tests the resulting variants without the conversion of information.
  - **White box fuzzing**. It generates target output based on complete knowl-edge and behaviour of the application. It repeats process of generating output as follows:
  - **Mutational black box fuzzing**. It generates the target output based on one or more seed inputs. It generates new inputs after random mutations

to ran dom locations (8). It is the extension to black box where random numbers are generated, but using random method.

- **Grey box fuzzing**. It generates target output based on minimal knowledge of the application (17). When specifications are very complex and sample data are easy to collect, data mutation is more appropriate than data generation. It gives lightweight feedback.

- **Web application scanners**. This is a quantitative way of doing software vulnerability analysis. This method is specifically used for finding the web applications vulnerabilities. It checks the volatility of the input for its correctness. Further, this can be analysed for white box or black box methodologies. White box testing analyses the source code manually. Black box testing checks the scanner fuzzing approach. This method is mainly applied in the testing stage of the system development. It comes out with a low false-positive ratio (10).

- **Static analysis techniques**. This is a stepwise execution method using quantitative way. It performs the activities to assess the input code, applies algorithms and generates output with expected vulnerabilities. It checks information flow's integrity and confidentiality (17). This can be further identified as false negative and false positive. False negative checks errors which are not yet being written, whereas false positive accesses only a subset of the required information. The scan ner can access only a subset of the required information. It determines whether a vulnerability exists or not.

- **Binary run-time integer-based vulnerability checker**. This is a quantitative method. It detects run-time data type-based vulnerability. It gives false-positive and false-negative results. It first converts the binary code to intermediate representa tion. It can be classified into categories as follows as per different data types (9).
  - **Integer overflow**. This may be due to overflow or underflow of data. Due to the limited range of integer variable with certain type, results may be of a larger or smaller value than expected. This vulnerability may affect the memory attachment.
  - **Signedness issues**. This issue may arise due to mathematical operation. Mathematical operation will be performed to get the overflow as per the different data type. They are likely to trigger integer wrapping. The sign may impact further steps of execution.
  - **String expansion**. Strings accessibility when working on characters can be problematic. Some characters are treated differently. These changes may create an impact on the workflow of the module. It may lead to changes in the targeted output.
  - **Format strings**. It handles changes in the string specifiers. They check the working of function w.r.t. the input. Depending upon format, different data types vulnerabilities can be added.
  - **Heap corruption**. Memory can generate vulnerabilities. Heap overflows can be triggered by memory allocation errors. It determines required buffer lengths to calculate the impact of overflow.

- **C Range Error Detector (CRED) approach**. It is a quantified method for the detection of software vulnerabilities. It handles the buffer overflow. It lacks the power to protect against all buffer overrun attacks. It breaks existing code and also produces too high overhead. CRED proved to be effective in detecting buffer overrun.
- **Module impact factor**. It checks the quality measurement of the modules w.r.t. the expected outcome. Higher module impact factor indicates an error with a high probability of causing damage. Wrong data from this module will cause more harm to the system. It is easier to generate abnormal data points. This impact factor will help find the quality parameter for the performance improvement w.r.t. complete application.
- **Buffer overflow rate**. This is a qualitative detection method. If the input data exceed the maximum amount of buffer overflow, it will overwrite and modify the value in the adjacent memory area. This rate is used to check the memory storage.
- **Module error tolerance rate**. It is a qualitative detection method to check the tolerance of accuracy. This is fault tolerant. This will give the tendency that the module will control the impact on other modules. This will generate the module-wise checking.

9.3.2 SOFTWARE VULNERABILITIES DETECTION TOOL

Figure 9.1 shows different detection tools used currently for the detection of software vul nerabilities. This figure shows the name of the tools used in the society.

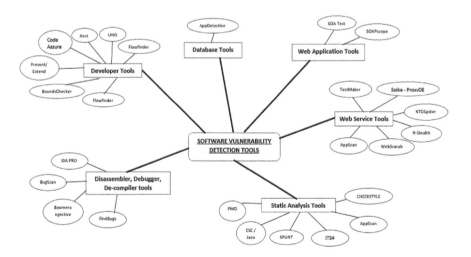

FIGURE 9.1 Software vulnerability detection tools.

9.4 MULTI-CRITERIA DECISION-MAKING (MCDM)

The decision of any problem can be taken based on the detailed analysis of the problem. Here, different parameters affecting the problem are analysed. These parameters can be classified as different criteria or alternatives to get the solution. The MCDM concept handles the process of making decision in the presence of multiple criteria or alternatives. It is an evaluation approach designed to deal with complexity. Here, alternative choices are analysed by considering a set of multiple and frequently conflicting criteria.

It solves the best alternative for the problem based on given set of selection criteria. It ranks the alternatives based on the mapping among the criteria to criteria. The highest ranked one is recommended as the best alternative to the decision-maker (19). MCDM techniques are commonly assessed qualitatively. MCDM entails uncertainty when the weighting process occurs subjectively based on the analyst experience (12). MCDM solution focuses on the constraints and preference on the priorities for the selection. It normalises the values assigned based on the comparison among the alternatives w.r.t. criteria. It takes normalisation w.r.t. expert knowledge.

9.4.1 LIST OF MCDM TECHNIQUES

The following are the MCDM methods studied that can be used for finding the software vulnerability analysis (21):

- **Multi-attribute utility theory**. It takes changes as per the uncertainty into account. This measures accuracy in quantitative manner. Criteria are nothing but working on the utility assigned to it. This utility is not a quality. Here, the calculated accuracy is convenient to measure. This method is extremely data intensive. Precise preferences need to be given to the criteria.

  Applications: Economic, financial, actuarial, water management, energy manage ment, agriculture.
- **Analytic hierarchy process (AHP)**. It uses pairwise comparisons among alternatives carefully in a hierarchical manner. It compares alternatives with respect to various criteria and estimates about criteria weights. It assesses various non-monetary criteria. Experts derive results based on priority scales. The AHP is designed for subjective judgements (13). A set of alternatives integrates hierarchical division by weighting the aspects considered in the analysis. It may empower the decision. The disadvantage of AHP is self-assessment bias affecting internal validity.

  Applications: Resource management, public policy, political strategy and planning.
- **Fuzzy set theory**. It focuses on use of cost-benefit analysis. It solves lots of problems related to imprecise and uncertain data. It handles rule-based analy sis. It tries to solve the problems with great complexity. The disadvantages of fuzzy set theory can sometimes be difficult to develop. It embraces vagueness.

  Applications: Engineering, economic, environmental, social, medical and man agement.
- **Case-based reasoning**. This requires extra knowledge of understanding. It improves the results over time when more cases are added to the database.

It is not good for its sensitivity to inconsistency in data. CBR is used in industries where previous cases are used as an experience.

Applications: Businesses, insurance, medicine and engineering designs.

- **Data envelopment analysis**. It uses a linear programming technique. It counts the relative effectiveness among alternatives. It gives the priorities to it. It checks for the most efficient alternative having a good rating. The efficiency can be analysed and quantified. It can uncover relationships that may be hidden. The disadvantage is that it does not deal with imprecise data. It assumes that all input and output data are known.

  Applications: Economic, medical, utilities, road safety, agriculture, retail and business problems.

- **Simple multi-attribute rating technique**. It handles multiple attributes for decision-makers. It requires two assumptions, i.e. utility independence and preferential independence. It converts weights into actual numbers. It requires less effort. It checks data with respect to each criterion. It is a complicated framework.

  Applications: Environmental, construction, transportation and logistics, military, manufacturing and assembly problems.

- **Goal programming**. It targets goals for set of data. As per the goal, the decision varies. It is able to choose the best from an infinite number of alternatives. It is not able to handle weight coefficients.

  Applications: Production planning, scheduling, healthcare, portfolio selection, distribution system design, energy planning and wildlife management.

- **ELECTRE (ELimination Et Choix Traduisant la REalit´e)**. This method is based on concordance analysis. It considers uncertainty and vagueness of data. Its pro cess and outcomes can be hard to explain. As preferences are incorporated, it will ignore the lowest performances under certain criteria. It cares for strengths and weaknesses of the alternatives. It is not directly identified. Results and im pacts need to be verified.

  Applications: Energy, economic, environmental, water management and trans portation problems.

- **VIKOR (from Serbian: VIseKriterijumskaOptimizacija I Kompromisno Resenje)**. In this method, the best alternative is selected by minimising regret group. It works on utility group theory. It compromises solution with an advantage rate. It is used in a highly complex environment. The performance rating is quantified as crisp values. It doesn't consider imprecise or ambiguous data (23).

- **PROMETHEE**. It performs several iterations to get the best results. It is an outranking method. It does not require the assumption about the criterion's proportionate. It does not provide a clear method to assign weights to each criterion.

  Applications: Environmental management, hydrology and water management, business and financial management, chemistry, logistics and transportation, man ufacturing and assembly, energy management and agriculture.

- **Simple additive weighting (SAW)**. It is a value function established based on a simple addition of scores. It represents the goal achievement under each criterion, which is multiplied by the particular weights. It has the ability to compensate for solution among set of criteria. The calculation is simple and can be performed without the help of complex computer programs.

The estimates by SAW do not always reflect the real situation. The result obtained may not be logical. The values of one particular criterion largely differ from those of other criteria.

Applications: Water management, business and financial management.

- **Technique for Order of Preference by Similarity to Ideal Solution**. It is an approach to identify an alternative that is closest to the ideal solution. It is farthest to the negative ideal solution. It gives multi-dimensional computing space. The number of steps of iteration remains the same irrespective of the number of attributes used in decision-making. The disadvantage of this method is that it uses Euclidean distance for calculations. It does not consider the correlation among the attributes.

    Applications: Supply chain management and logistics, design, engineering and manufacturing systems and business and marketing.

- **Analytic network process (ANP)**. It handles hierarchical alternatives at the lower levels. It checks linear log with the goal at the top. The dependency among the criteria is not required. Prediction is accurate based on the priorities given by the feedback. This MCDM method is similar to AHP. The elements of the same cluster are compared among themselves without checking the hierarchy. The level of each element may dominate and may get dominated in pairwise comparisons.

- **The weighted sum model (WSM)**. It is simple mathematical calculation. It work on single dimension. It varies across the range across criteria (7). It is useful for evaluating several alternatives in accordance with different criteria. They are expressed in the same units of measurement. It gives relative order of magnitude for standardised scores. It is uncomfortable on multi-dimensional problems.

- **The weighted product model (WPM)**. Similar to WSM, it performs easy calculation. It uses relative values instead of actual ones. It compares criteria with others by the weights. It also checks ratio for each criterion. If the number of alternatives is large, then it's lengthier. It will be more difficult to solve. No solution will be available if equal weights are assigned to decision matrices (22).

Figure 9.2 shows the hierarchical structure of different MCDM techniques used. This tree structure shows the inheritance of properties from top level to down level.

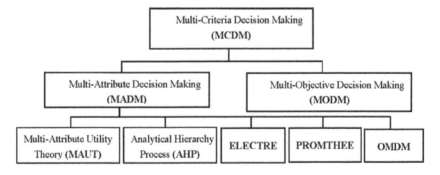

FIGURE 9.2   List of Multi-Criteria Decision Making Methods.

9.4.2 NOTATIONS USED IN MCDM

- **Alternatives**. Alternatives are the different choices of action available. Usually, alternatives are finite in number. They are supposed to be evaluated for its importance and then eventually ranked.
- **Multiple attributes**. Attributes can be considered as goals or decision criteria. Attributes with alternatives represent different dimensions. Attributes may have major attributes and then hierarchy of set of attributes. Depending on the analysis of problem, hierarchy can be further enhanced. This set of attributes helps to get the accurate results.
- **Conflict among attributes**. As different attributes represent different dimensions of the alternatives, they may conflict with each other. This conflict may help to decide priority among the attributes. This also helps to decide the relationship among the attributes.
- **Incommensurable units**. Attributes may be with different units of measure. This differentiation may be difficult to solve. Data standardisation is required to solve this problem. As per the problem statements, set of rules can be applied for measurement.
- **Decision weights**. Weights are normalised before assigning to the alternatives. This set of matrix helps to calculate performance index.
- **Decision matrix**. An MADM problem can be easily expressed in matrix format. A decision matrix A is an $(M \times N)$ matrix, in which element a_{ij} indicates the performance of alternative and A_i indicates evaluation in terms of decision criterion C_j (for $i = 1,2,3,\dots, M$ and $j = 1,2,3,\dots, N$). Experts determine the weights of relative performance of the decision criteria denoted as W_j, for $j = 1,2,3,\dots, N$ using the formula:

$$x_{ij} = \frac{a_{ij}}{\sqrt{\sum_{i=1}^{M} a_{ij}^2}} \tag{9.1}$$

Therefore, the normalised matrix X is defined as follows:

$$
\begin{bmatrix}
x_{11} & x_{12} & \cdots & x_{1N} \\
x_{21} & x_{22} & \cdots & x_{2N} \\
\cdot & \cdot & \cdots & \cdot \\
\cdot & \cdot & \cdots & \cdot \\
x_{M1} & x_{M2} & \cdots & x_{MN}
\end{bmatrix}
$$

where M is the number of alternatives, N is the number of criteria, and x_{ij} is the preference measure of the ith alternative with respect to j-th criterion.

9.4.3 IMPORTANT STEPS USED IN MCDM MODELS TO OBTAIN THE RANKING OF ALTERNATIVES

- **Determine criteria and alternatives**. Understand the problem. The decision-maker lists the different criteria and lists the best alternatives as per criteria. The rela tion between the criteria and alternatives is developed. It builds the relevancy among them (20).
- **Develop decision criteria**. Based on the understanding of the problem, experts will develop the important criteria for the final solution. This gives the priorities among the criteria. Complex criterion can be further subdivided into simpler criteria for analysis.
- **Allocate the weight to the criteria**. Based on the expert knowledge, weights are assigned to the criteria. Weights are also evaluated for each criterion w.r.t. alternatives. Depending on the MCDM techniques, weighting criteria are different.
- **Develop and analyse the alternatives**: Using any MCDM techniques, find the highest possible alternative as the solution. Analyse the solution w.r.t. the specified problem. The relative importance of the criteria is calculated based on the expert knowledge. Impacts of the alternatives in relation to the criteria are evaluated.
- **Select and implement alternatives**. Process the numerical values to determine each criterion with respect to the alternatives. Alternatives are ranked based on MCDM techniques. It relates the relation between the criteria to available alternatives and finds the priority-based selection for combination.
- **Evaluate the result**. Highest or lowest ranked alternatives are selected as final alternatives. This selection is totally based on the techniques that are used in relation to criteria. Results are evaluated for the selection of best alternatives.

Figure 9.3 shows the general flow chart of multi criteria based decision making concepts. These general steps are further varied as per different techniques.

9.5 ANALYSIS OF SOFTWARE VULNERABILITIES USING MCDM

To make decision using MCDM, the following mathematical model can be used. Let m of vulnerabilities to be assessed and prioritised a n numbers of decision criteria as vulnerability detection techniques. The vulnerabilities are denoted as V_i (for $i = 1,2,3,\ldots, m$) and the criteria as C_j (for $j = 1,2,3,\ldots, n$). Each criterion is associated with a weight, denoted as W_j (for $j = 1,2,3,\ldots, n$). The higher the weight is, the more important the criterion is assumed to be. The final performance index is calculated based on the sum of the weights w.r.t. input $[\sum_{j=1}^{n} W_i = 1]$. The final weights are best summarised in a decision matrix. The corresponding quantitative score P_{Vi} of each Vi is given by the equation: $[Pv_i = \prod_{j=1}^{n} (a_{ij})W^j]$.

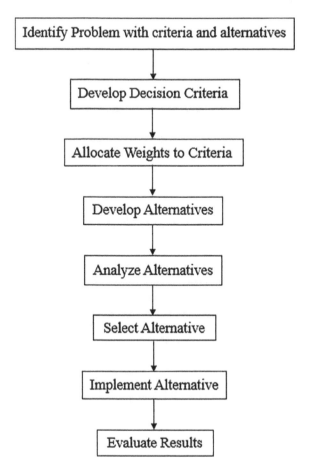

FIGURE 9.3 Flow chart of the general MCDM method.

This section focuses on finding the best alternatives as a solution using the AHP, SAW, WPM and WSM. Abbreviation used in the mathematical descriptions are Insecure interaction between components (II), Risky analysis (RA), Missing/ broken access (MA), Cross-Site Scripting (CS), known vulnerable component (KN) and Design vulnerabilities (DV) as shown in the Figure 9.4.

9.5.1 SOLUTION USING ANALYTIC HIERARCHY PROCESS (AHP)

The analytic hierarchy process (AHP) is a MCDM system. It is used to solve complex decision-making problems. The AHP is implemented in the software of experts' choice. The steps of execution are as follows (27):

- **Under problem situation**. Define the problem, determine the criteria, and identify the alternatives. Software vulnerabilities are listed as criteria. Here, different detection methods are identified as alternatives.

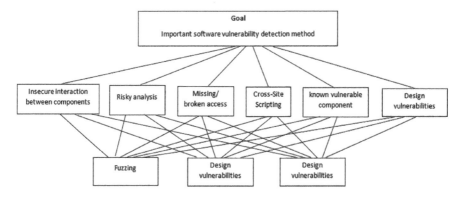

FIGURE 9.4   Problem statement.

- **Hierarchical structure of problem characteristics**. Structure a hierarchy of alternatives. Make pairwise comparisons among criteria w.r.t. alternatives. Rate the relative importance between each pair of decision. Here, a detailed analysis of the problem statements is done. After doing the prioritisation of these vulnerabilities as criterion, modelling can also be done on these critical vulnerabilities. As shown in Table 9.1, AHP uses 1-9 scale for the prioritisation process.

Intermediate numerical ratings of 2, 4, 6 and 8 can be assigned, someone could not decide whether one criterion (or alternative) is moderately more important than the other one.

- Synthesise the results to determine the best alternative. The output of AHP is the set of priorities of the alternatives (Table 9.2).
- The second step for a detailed comparison about the criteria with respect to the selec tion of the method of MCDM is given by the following things. (Table 9.3)
- Check the impact of software venerability w. r. t. selection of MCDM techniques. Synthesis the results by doing composition of the impacts. (Table 9.4)

TABLE 9.1
AHP Scale

Numerical Ratings	Verbal Judgements
1	Equally important (preferred)
3	Moderately more important
5	Strongly more important
7	Very strongly more important
9	Extremely more important

TABLE 9.2
Pairwise Comparisons for Software Vulnerabilities

	II	RA	MA	CS	KN	DV	Weight
Fuzzing	1	4	3	1	3	4	0.32
Design vulnerabilities	1/4	1	7	3	1/5	1	0.14
Binary run-time integer-based vulnerability checker	1/3	1/7	1	1/5	1/5	1/6	0.03

TABLE 9.3
Comparison of Vulnerabilities w.r.t. Techniques

	II				RA				MB			
	F	W	B	wt	F	W	B	wt	F	W	B	wt
F	1	1/3	1/2	0.16	F	1	1	1	0.33	F	1	5	5	0.45
W	3	1	3	0.59	W	1	1	1	0.33	W	1/5	1	1/5	0.09
B	2	1/3	1	0.25	C	1	1	1	0.33	B	1	5	1	0.46

	CS				KN				DV			
	F	W	B	wt	F	W	B	wt	F	W	B	wt
F	1	9	7	0.77	F	1	1/2	1	0.25	F	1	6	4	0.69
W	1/9	1	1/5	0.05	W	2	1	2	0.50	W	1/6	1	1/3	0.09
B	1/7	5	1	0.17	C	1	1/2	1	0.25	B	1/4	3	1	0.22

TABLE 9.4
Comparisons and Synthesis of Software Vulnerabilities

	0.32	0.14	0.03	0.13	0.24	0.14	Composite
	II	RA	MA	CS	KN	DV	Impact
F	0.16	0.33	0.45	0.77	0.25	0.69	0.32
W	0.59	0.33	0.09	0.05	0.5	0.09	0.38
B	0.25	0.33	0.46	0.17	0.25	0.22	0.25

9.5.2 SIMPLE ADDITIVE WEIGHTING METHOD

This method is simple to implement. It follows the general flow of MCDM concepts. It calculates the performance index based on the additive weighting calculations. For simplicity, calculated weights from the previous method are used to find best alternatives in this method as well. Following are the steps of executions.

- Each alternative is assessed w.r.t. each criterion. Overall performance score is evaluated using $p_i = \sum_{j=1}^{M} W_j * M_{ij}$

  M_{ij} = The measure of performance of alternative i w.r.t. attribute j
  w_j = The weights of alternatives.

TABLE 9.5
SAW/WSM Method

Criteria	II	RA	MA	CS	KN	DV
F	0.16	0.33	0.45	0.77	0.25	0.69
W	0.59	0.33	0.09	0.05	0.5	0.09
B	0.25	0.33	0.46	0.17	0.25	0.22
Weights	0.32	0.1	0.09	0.27	0.15	0.07
Beneficial (+) Non-beneficial (−)	(+)	(+)	(−)	(+)	(+)	(−)
Calculated value	0.59	0.33	0.09	0.77	0.5	0.09

- The ratio is evaluated for beneficial and non-beneficial values. Beneficial values indicate the high impact value, whereas it is vice versa for non-beneficial values.
- It is a proportional linear transformation of raw data, which means the relative order of magnitude of standard, and the score remains equal (Table 9.5).
- For beneficial attributes, its higher values are calculated using the formula:

$$w_{ij} = w_{ij}/w\text{max}.$$

- For non-beneficial attributes (Table 9.6), its lower values are calculated using the formula:

$$w_{ij} = w_{\text{min}}/w_{ij}.$$

- The performance index for each criterion is calculated using the formula:

$$p_{i/j} = \left[\sum_{j=1}^{M} w_j * M_{ij}\text{normal}\right] \Big/ \left[\sum_{j=1}^{M} w_j\right].$$

For $F = P(F/W) = (0.16/0.59)0.32 + (0.33/0.33)0.1 + (0.45/0.09)0.09 + (0.77/0.05)0.27 + (0.25/0.5)0.15 = 6.96$.
$P(W/B) = 6.45 \ P(F/B) = 5.94$.

Above calculation for the same problem, shows that fuzzing is the best alternativeto solve the six software vulnerabilities as compared to other alternatives.

TABLE 9.6
SAW/WSM Method

Criteria	II	RA	MA	CS	KN	DV
F	0.271	1	0.2	1	0.5	0.13
W	0.423	1	1	0.064	1	1
B	0.423	1	0.195	0.22	0.5	0.4

TABLE 9.7

WPM Method

Criteria	II	RA	MA	CS	KN	DV
F	0.16	0.33	0.45	0.77	0.25	0.69
W	0.59	0.33	0.09	0.05	0.5	0.09
B	0.25	0.33	0.46	0.17	0.25	0.22
Weights	0.32	0.1	0.09	0.27	0.15	0.07
Beneficial (+) Non-beneficial (−)	(+)	(+)	(−)	(+)	(+)	(−)
Calculated value	0.59	0.33	0.09	0.77	0.5	0.09

9.5.3 Weighted Product Model

This method follows the same execution steps as that of AHP specified above. It checks the performance index based on the product of weights assigned. For simplicity, calculated weights from AHP techniques are used. Following are the steps of execution.

- Each alternative is assessed w.r.t. to each criterion. The overall performance score is evaluated using the performance index formula:

$$p(A_k \ / \ A_i) = \prod_{i=1}^{n} (a_{kj} \ | \ a_{ij})^{wj} \tag{9.2}$$

- Calculation considering the beneficial and non-beneficial parameters needs to be performed. Considering the previous example final table for evaluation will be as follows (Table 9.7).

  $P(F/W) = (0.16/0.59)^{0.32} + (0.33/0.33)^{0.1} + (0.45/0.09)^{0.09} + (0.77/0.05)^{0.27} + (0.25/0.5)^{0.15} + (0.69/0.09)^{0.07} = 6.96 \ P(W/B) = 6.45.$
  $P(F/B) = 5.94.$

- The WPM is used in the similar track as that of simple additive method. The first criterion is the best criterion.

9.6 OUTCOME FROM THE MATHEMATICAL MODEL

It is observed from the mathematical model that the best alternatives are selected based on the comparative analysis. The relation between criteria and alternatives is analysed. It is observed that fuzzing is the best method to detect the vulnerability. Further, if the mathematical model is normalised with a detailed structure of vulnerabilities, we can find mapping w.r.t. fuzzing techniques as well. Hence, more accurate methods can be analysed.

9.7 CONCLUSIONS

Automated vulnerability discovery is a game between adversaries. Understanding the working and impact of different vulnerabilities helps to learn more software issues. To understand the nature and distribution of security vulnerabilities in source code, the type of information is usually not available in an executable format. Detection methods help to understand software vulnerability. MCDM techniques allow to understand the problem in detail. It understands the hierarchical representation of the problem. It calculates the relation between criteria and solution. It builds the relation between them. Based on the criteria index important criteria can be calculated. Hence based on this performance index of the overall alternatives are calculated. Finding the best alternatives is the objective of this chapter. To achieve the goal, a detailed analysis of the problems in terms of vulnerabilities with their impact was performed. In the same manner, different detection methods are analysed w.r.t. vulnerabilities evaluated as alternatives. MCDM techniques allow to find the best alternatives.

REFERENCES

[1] H. Ghani, J. Luna, N. Suri. 2013. Quantitative assessment of software vulnerabilities based on economic-driven security metrics. *International Conference on Risks and Security of Internet and Systems (CRiSIS)*.

[2] R. Amankwah, P. K. Kudjo, S. Y. Antwi. 2017. Evaluation of software vulnerability detection methods and tools: A review. *International Journal of Computer Applications* 169, 8.

[3] J. Willy, M. Amel, C. Ana. 2010. *Software Vulnerabilities, Prevention and Detection Methods: A Review.*

[4] P. Zeng, G. Lin, L. Pan, Y. Tai and J. Zhang. 2020. Software vulnerability analysis and discovery using deep learning techniques: a survey. In *IEEE Access*, 8, pp. 197158–197172, https://doi.org/10.1109/ACCESS.2020.3034766.

[5] B. Liu, L. Shi, Z. Cai, M. Li. 2012. Software vulnerability discovery techniques: A survey. *Fourth International Conference on Multimedia Information Networking and Security.* 978-0-7695-4852-4/12 $26.00. © 2012 IEEE. https://doi.org/10.1109/MINES.2012.202.

[6] P. Godefroid, A. Kiezun, M. Y. Levin. 2008. Grammar-based whitebox fuzzing. *Proceedings of the 2008 ACM SIGPLAN Conference on Programming Language Design and Implementation.*

[7] M. Bohme, C. Cadar, A. Roychoudhury. 2020. *Fuzzing: Challenges and Reflections.* National University of Singapore, Singapore IEEE Software Published by the IEEE Computer Society, IEEE.

[8] M. Woo, S. K. Cha, S. Gottlieb, D. Brumley. 2013. Scheduling black-box mutational fuzzing. *Proceedings of the 2013 ACM SIGSAC Conference on Computer & Communications Security.* https://doi.org/10.1145/2508859.2516736.

[9] T. Clarke. 2009. *Fuzzing for Software Vulnerability Discovery. Technical Report RHUL-MA-2009-04.* Department of Mathematics Royal Holloway, University of London.

[10] G. Vache. 2009. Vulnerability analysis for a quantitative security evaluation, *3rd International Symposium on Empirical Software Engineering and Measurement, 2009*, pp. 526–534, https://doi.org/10.1109/ESEM.2009.5315969.

[11] O. H. Alhazmi and Y. K. Malaiya, 2005. Quantitative vulnerability assessment of systems software. *Annual Reliability and Maintainability Symposium, 2005 Proceedings.* pp. 615–620, https://doi.org/10.1109/RAMS.2005.1408432.

[12] de Azevedo Reis, G., de Souza Filho, F.A., Nelson, D.R. et al. 2020. Development of a drought vulnerability index using MCDM and GIS: study case in Sao Paulo and Ceara, Brazil. *Natural Hazards* 104, 1781–1799. https://doi.org/10.1007/s11069-020-04247-7.

[13] N. Liu, J. Zhang, H. Zhang and W. Liu. 2010. Security assessment for communication networks of power control systems using attack graph and MCDM. *IEEE Transactions on Power Delivery* 25, 3, 1492–1500, https://doi.org/10.1109/TPWRD.2009.2033930.

[14] E. Kornyshova and C. Salinesi. 2007. MCDM techniques selection approaches: State of the art. *2007 IEEE Symposium on Computational Intelligence in Multi-Criteria Decision- Making*, pp. 22–29, https://doi.org/10.1109/MCDM.2007.369412.

[15] M. Velasquez, Patrick T. 2013. An analysis of multi-criteria decision making methods. *International Journal of Operations Research* 10, 2, 5666.

[16] C. Luo, W. Bo, H. Kun, L. Yuesheng. 2020. Study on software vulnerability characteristics and its identification method. *Mathematical Problems in Engineering* 2020, Article ID 1583132, 6 p. https://doi.org/10.1155/2020/1583132.

[17] B. Liu, L. Shi, Z. Cai, M. Li. 2012. Software vulnerability discovery techniques: a survey. *Fourth International Conference on Multimedia Information Networking and Security.*

[18] L. Ping, S. Jin, Y. Xinfeng. 2011. Research on software security vulnerability detection technology. *International Conference on Computer Science and Network Technology.*

[19] Yurdakul, M., Tansel, I. C. Y. 2009. Application of correlation test to criteria selection for multicriteria decision making (MCDM) models. *The International Journal of Advanced Manufacturing Technology* 40, 403–412. https://doi.org/10.1007/s00170-007-1324-1.

[20] Arslan M. C. 2002. *A Decision Support System for Machine Tool Selection*, M.S. Thesis, Sabancı University.

[21] M. Velasquez, P. T. Hester. 2013. An analysis of multi-criteria decision making methods. *International Journal of Operations Research* 10, 2, 5666. M. Aruldoss, T. Miranda Lakshmi, V. Prasanna Venkatesan, 2013. A survey on multi criteria decision making methods and its applications. *American Journal of Information Systems* 1, 1, 31–43. https://doi.org/10.12691-ajis-1-1-5.

[22] U. S. Kashid, A. U. Kashid, S. N. Mehta. A review of mathematical multi-criteria decision models with a case study. *International Conference on Efficacy of Software Tools for Mathematical Modeling (ICESTMM'19).*

[23] S. Opricovic, G.-H. Tzeng. 2004. Compromise solution by MCDM methods: A comparative analysis of VIKOR and TOPSIS. *European Journal of Operational Research* 156, 445–455.

[24] S. Narang, P. K. Kapur, D. Damodaran and R. Majumdar. 2018. Prioritizing types of vulnerability on the basis of their severity in multi-version software systems using DEMATEL technique, *2018 7th International Conference on Reliability, Infocom Technologies and Optimization (Trends and Future Directions) (ICRITO)*, pp. 162–167, https://doi.org/10.1109/ICRITO.2018.8748720.

[25] H.A. Proper, D. Greefhorst. 2011. *The Roles of Principles in Enterprise Architecture.* https://doi.org/10.1007/978-3-642-20279-7-2.

[26] Kotusev, Svyatoslav. 2019. Enterprise architecture and enterprise architecture artifacts: Questioning the old concept in light of new findings. *Journal of Information Technology* 34, 102–128. https://doi.org/10.1177/0268396218816273.

[27] A. M. Ghaleb, H. Kaid, A. Alsamhan, S. H. Mian, and L. Hidri. 2020. Assessment and comparison of various MCDM approaches in the selection of manufacturing process. *Advances in Materials Science and Engineering*, Article ID 4039253, 16 p. https://doi.org/10.1155/2020/4039253.

10 On a Safety Evaluation of Artificial Intelligence-Based Systems to Software Reliability

Sanjay Kumar Suman
St. Martin's Engineering College

L. Bhagyalakshmi
Rajalakshmi Engineering College

Rajeev Shrivastava
Princeton Institute of Engineering &
Technology for Women

Himanshu Shekhar
Hindustan Institute of Technology and Science

CONTENTS

DOI: 10.1201/9780367816414-10

10.1   INTRODUCTION

In the past years, the popularity of artificial intelligence (AI) system has become more than increasing application number reported. Examples are given below.

- Process of data
- Assistant systems
- Voice, face, speech recognition.

'The application of AI relevant most safety'. It requires safety assessment, 'apply consequence for functional safety assessment'. In this book chapter, we have considered the safety assessment of AI systems.

In the second part, the definition of AI system is given. In the third part, how to show and obtain safety integrity level in AI systems is explained. In the fourth part, deeper knowledge and view about AI systems is presented. It is so necessary for AI systems to understand about an approach in terms of safety functions. In the fifth part, software reliability and safety of AI systems is described. In the last part, conclusions are drawn.

10.2   WHAT IS ARTIFICIAL INTELLIGENCE?

Dartmouth College used artificial intelligence in 1956. Many concepts were proposed by researchers. AI is defined as intelligence demonstration by machine. Through this type of technique, cognitive problem solving, learning and functions can be done. 'There are some criteria follow point for AI or not'.

Using speech system.

- Consciousness system.
- System of self-awareness.

However, the outcomes are genuinely remarkable. Many articles concerning deep learning have been presented, for example Hättasch and Geisler (2019). 'In AI applications, there are complete … as we know'. Some approaches have recently been made in terms of safety; take a look at the proposed UL 4600 standard (2019). It necessitates a safety strategy, with AI algorithms being used in some circumstances for autonomous vehicles. UL 4600 additionally explains not just what is being argued, but also how it is not being argued. "Conformance with this standard is not a guarantee of a safe automated vehicle", as written in the preamble. Rather than "repeatable assessment of the thoroughness of a safety case", its importance is on "repeatable assessment of the thoroughness of a safety case". UL 4600 is a safety standard used to intend the extension of IEC 61508. Some standard committees, for example German DKE, processed and focused on lifecycle-oriented approaches. A λ AI measure Putzer (2019) propagates similar to give a succinct definition of danger rate in functional safety.

10.3   DOES ARTIFICIAL INTELLIGENT REQUIRE A SIL?

In this section, we'll have a look at the level of safety integrity of AI system; if yes, then how to determine. The preface of SIL system is used for standardisation of the

Situation with other E/E/PE systems

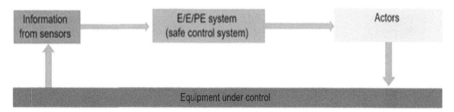

FIGURE 10.1 The E/E/PE controlling system.

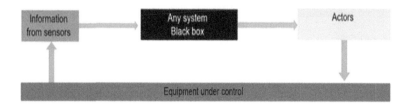

FIGURE 10.2 Controlling system with an arbitrary black box.

safety function. The mother standardisation is also known as IEC 61508. 'The determination about SIL system … reader'.

Figure 10.1 shows a normal situation about electronic, electric, programmable electronic system (E/E/PE system). 'It has under equipment control system…operate actor systems'.It depends on consequences failure behaviour of this controlling system to get level of safety integrity (SIL).

It is no longer the case that we have a controlling system by analysing hazards and determining the SIL system. In any case, it considers black box; Figure 10.2 depicts it.

Nowadays, the AI system also is a black box system. 'It is also safety integrity level ... by the E/E/PE methods'.

'SILs assessment rule only different types … implement black box system'.

What so AI applications expect from different SILs? The failure consequence mainly depends on possibility risk as follows:

- Process of data – it depends on the result.
- System assistance – 'SIL system normally … this type of system' is OK.
- Speech, face, voice recognition – it depends on the result, whether safe backup or activated result.

'In this hazard cases risk analysed … relevant safety apply for this'.It is not determine and compulsory for one. Standard function relevant safety apply for this.

10.4 LOOKING INSIDE AI

Architecture of AI: A simple architecture of AI is shown in Figure 10.3. This type of architecture is inspired by Wand (2017), although it does not favour it.

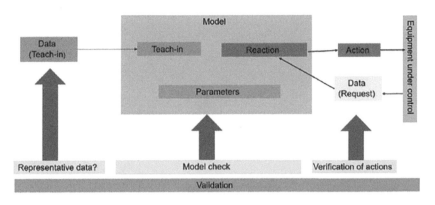

FIGURE 10.3   An AI architecture system.

Inside this AI system's models, many key features are present. This is a very flexible model.

'It needs to teach … for circumstances in the future'.

It is a very compulsory situation to avoid. It is mentioned as an example by Corni (2019).

Racism was detected in an AI system learning for imported collection of data via non-representation. Parameters are set after learning in this model.

'Requesting data and activation actor control … is used later'.

'The possibility to continue teaches … into exploitation'. It has some importance as:

- Model checking.
- Representativeness data checking.
- Verification of data.

In the following, we have a deeper look into many parts of AI system. Figure 10.3 shows a refined architecture of an AI system.

10.5   SOFTWARE RELIABILITY

'To improve the reliability…to target the range of services'.

'If software is in form of mission-critical…for market-ready'.

Failure of the Software:
'The use of software reliability … due to software failure'.

'It is not failure behaviour … the hardware'.

'Computer present the software function … the computer statements'.

If software is a failure, it is because of either implementation or design errors. Design errors show wrong assumptions about computer system operation. Implementation errors show confusion symbols.

Software Reliability: 'In the instance of hardware…failure model for future'. Many organisations are no longer testing the software.

'Management always surprise cost … in the favour of designing activity'.

'A new feature adding … for testing features exist'. A good quality of software introduces developing cycle.

- 'Advance preparation and management program'.More effort for testing program.
- To allow scheduling and budget, it covers testing requirements.

   Software reliability engineers have extensive knowledge of all stages and duties associated with a comprehensive software reliability programme. Leading and supporting reliabilities of program as:
   - Reliability of the program
   - Allocation of the program
   - Define operational profile
   - Analysis of program
   - Testing of the program
   - Planning of the program

Allocation of Software Reliability:

'It is used for define task … software items'.

It may be a hardware/software system. It may be related to the independent application of the software. It is a stand-alone relay of the program.

'In this case, our goal is to bring reliability system … cost constraints'.

i. It follows some tasks to assist with our organisation:
   - Requirement of software reliability is derived.
   - Optimisation of reliability.
   - Scheduling of software reliability.
   - Cost based on our constraints.
   - Costs depend on our goal of software reliability.

ii. **Defining and Analysing Operational Profiles**: Software liableness is inextricably linked to how an application is used in the field, far more so than hardware reliability. Only if a software flaw occurs while the system is in use may it result in a system failure. An issue that is not addressed in a certain operational mode will not result in any failures. If it's in code that's part of a commonly used "operation", it'll fail more frequently. As a result, in software reliability engineering (SRE), we concentrate on the software's operational profile, which weighs the likelihood of each operation occurring. Unless safety constraints necessitate a change, we shall prioritise our testing in accordance with this profile.

iii. **Software Reliability Models**: Reliability models, particularly reliability growth models, are frequently associated with SRE. When used correctly, these models are effective in providing assistance for management decisions such as:
   - Scheduling for testing
   - Allocation of resources
   - Marketing time

- Resource maintenance allocation
- 'To adjust the growth rate … to enable rate of failure'. The term "software reliability" specifies the ability of software to meet the requirement of its users. It is described as the likelihood that software will operate without error over a given amount of time. The term "failure" refers to a situation in which the program did not perform as expected by the customer. This broad delineation of failure ensures that the concept of dependability embraces the majority of quality qualities such as accuracy, performance adequacy and usability. The term "reliability" refers to a user-centred approach to software quality that emphasises how well the product really performs. Alternative perspectives on software quality are introspective, developer-oriented viewpoints that link product quality to its "complexity" or "structure". Fortunately, software liableness is not only one of the paramount and instant criteria of software quality, but also one of the most straightforward to define and assess. Software liableness is a scientific field that necessitates the use of exact language. The two most significant words are "failure" and "fault". When the program's outer behaviour departs from the user's expectations, this is referred to as a software failure. A software defect is a flaw in a program that causes it to fail when run under certain conditions. It is sometimes referred to as a "bug". Two major factors influence the likelihood of failure:
- The amount of faults in the programme being used; the higher the number of bugs, the higher the number of failures;
- The conditions in which it is being executed (also known as the "operational profile"). Some conditions may be more difficult than others, resulting in a higher number of failures.

The possibility of a computer program's failure-free operation over a certain amount of time under specified conditions is known as software reliability. A secretary's text-editor, for example, may have a reliability of 0.97 for 8 hours; a hacker's text editor, on the other hand, may only have an 8-hour reliability of 0.83. Consider the following scenarios to obtain a better grasp of the flight-critical aircraft systems' dependability needs.

i. **Reliability**: $R(t)$ stands for reliability, which is the chance of failure-free operation for an extended period of time t.

ii. **Probability of Failure**: The probability that the software will fail before time t is represented by $F(t)$. There is a link between reliability and the likelihood of failure.

$$R(t) = 1 - F(t).$$

iii. **Failure Density**: The probability density for failure at time t is given by $f(t)$. It has something to do with the likelihood of failure $f(t) = d / dt F(t)$.
    In the half-open interval $(t, t+t]$, the likelihood of failure is $f(t) \cdot \delta t$.

iv. **Hazard Percentage**: The conditional failure density at time t, indicated by $z(t)$, is the failure density if there had been no failures up to that moment. In other words,

$$z(t) = f(t) / R(t).$$

The hazard rate and reliability are linked by

$$R(t) = e^{-\int_0^t z(x)\, dx}.$$

When the hazard rate is constant, there is a significant exception ϕ. The failure density follows an exponential curve in this scenario.

$f(t) = \phi e^{-\phi t}$. The likelihood of failure is given by $F(t) = 1 - e^{-\phi t}$, and the dependability is provided by $R(t) = e^{-\phi t}$.

v. **The Mean Value Function**: (μt) denotes the average. By time t, the number of failures has increased.

vi. **Failure Intensity**: (λt) denotes the number of failures per unit of time at time t. This is related to the mean value function via analogy $\lambda(t) = d/dt\mu(t)$. The estimated number of failures in the half-open interval $(t, t + \delta t]$ is $\lambda(t) \cdot \delta t$.

The most popular metric for measuring software reliability is failure intensity. The numbers connected with reliability are random variables, and the reliability models are based on the mathematics of random or stochastic processes, due to the complexity of the elements causing the incidence of a failure. Because failure is typically followed by another failure, the number of mistakes in a program, as well as the probability distributions of the dependability model's components, change with time. Reliability models, to put it another way, are based on non-homogeneous random processes. A variety of software dependability models have been developed. The "Basic Execution Time Model" is the most accurate and commonly recommended model. This model's development is detailed below.

i. **Execution of the Basic Modelling**: In this derivation, the software failure process is represented as a non-homogeneous Poisson process (NHPP), a type of Markov model (and that of most other reliability models). The total number of failures at time t is represented by $M(t)$. Assumptions are made as follows:
1. By time 0, $M(0) = 0$; there have been no failures.
2. The process has independent increments, which means that the value of $M(t + \delta t)$ is determined only by the current value of $M(t)$ and is unaffected by the process's past.

10.6 SOFTWARE RELIABILITY DISCUSSION

Software dependability modelling is a significant scientific endeavour. Many businesses have a vested interest in the long-term viability of their software products such as embedded system makers (where maintenance can be difficult), and those with highly stringent dependability requirements have pursued it with passion. The "Basic Execution Time Model" presented here has been certified across a number of significant projects and has the advantage of being simple to use. 'Being quite simple when compared to many other models'.The "logarithmic Poisson concept", a comparable model, has received less attention, yet may be preferable in some cases. The time base for each of these models is execution time. One of the key points for their better accuracy over prior models that depended on man-hours or other human-oriented time is that they don't use man-hours or other human-oriented time. 'They don't use man-hours or other human-oriented time'. Metrics is because of this. It's hardly unexpected that execution time should improve: The frequency of failures should, after all, be largely influenced by the amount of exercise the software has got. Musa et al. illustrated how to transform from a machine-oriented time perspective to a human-oriented time perspective, which is generally required for the model's results to be applied. A number of problems may make the application of dependability models more difficult. The basis for prediction is the collection of accurate failure data early in the project's life cycle, whereas reliability is concerned with counting failures. If any of the following criteria exist, the data may be untrustworthy, and forecasts based on it may be incorrect:

- There is a lack of clarity or understanding about what defines a setback.
- There is a significant change in the operational profile between the data collection (e.g. testing) phase and the working phase, or
- The software is up to date always changing and evolving.

We highlight that many AI software development situations are similar to the ones that make reliability modelling difficult to apply. In the second section of the report, we will return to this topic.

10.7   CHARACTERISTICS OF AI SOFTWARE

The application of the quality management system will be discussed in this section of the report.

In Part I, quality assurance procedures and metrics were applied to AI software, and now, we'll move on to Part II to have a look at a couple of solutions designed specifically for this purpose. 'This kind of computer software'. We are currently confronted with a problem in that the definition of AI software is unclear; in fact, AI practitioners disagree on what constitutes AI. Parnas distinguishes between two modern AI concepts, AI-1 and AI-2: AI-1 is a problem-solving paradigm in which computers are used to answer that could previously only be addressed by human intelligence. 'AI-2 is a problem-solving paradigm ... by human intelligence'. AI-2 is a technique-oriented notion that links artificial intelligence to the use of specific programming approaches, particularly heuristic-based ones, as well as the explicit representation of "knowledge". These concepts aren't mutually exclusive; in fact, most AI software incorporates AI-1 and AI-2 components, making AI software SQA more difficult. The issues that AI-1 attempts to solve are frequently ill-defined, and the job that the programme is supposed to do is rarely specified in detail. As a result, the labels "success" and "failure" are vague, making evaluation challenging. Furthermore, AI-2 is sensitive and unstable because of the heuristic processes used: Using very similar inputs, you could get very different results.

This makes extrapolating from test case behaviour extremely dangerous. For the purposes of this report, we'll focus on AI applications and methodologies. This could be applied to the development of civil aviation's "intelligent cockpit assistance". BBN examines the prospect of such uses in a NASA Contractor Report. In general, they believe that "expert systems" for monitoring and diagnosing faults, as well as "planning assistants" for assistance with planning, topics such as fuel and thrust control, would be the most effective aids. (Reiter provides a basic explanation of defect detection, but Georgeff provides a good overview of planning.) Natural language voice recognition and creation, and the human aspects of integrating such aids into the cockpit are other significant AI technologies for this application. As our AI software paradigms, expert systems and, to a lesser extent, planning systems will be chosen because they have been identified as being of particular importance as intelligent cockpit aids, as well as their important place in current AI applications in general. Since there are currently no intelligent cockpit aids, we'll look at expert and planning systems in general, but we'll pay specific attention to concerns such as fault monitoring and diagnostics where applicable. The NASA Langley Research Center prototype defect monitoring system is an excellent one. This is a sample of the type of technology that could be one of the first AI systems placed on commercial planes. We're looking at "knowledge-based" AI, which means it has an explicit evident of knowledge about some aspects of the outside world. Knowledge is implicitly incorporated in traditional software in the form of algorithms or procedures. The programmer knows how the computer will handle payrolls or radar photographs, and he or she saves this information as processes (pre-planned sequences of activities), thus the phrase "procedural knowledge". In traditional software, knowledge is implicitly represented, whereas information-based software incorporates both a reasoning component that may use such knowledge to solve problems and an explicit declarative representation of information. For example, a knowledge-based system that transforms temperatures between Celsius and Fahrenheit could store information about the situation in the declaration.

$$C = (F - 32) \times 5 / 9.$$

This single declaration, when combined with a constraint-fulfilling reasoning component, would allow the system to convert Fahrenheit to Celsius and vice versa. In contrast, a traditional system would encode these data in the following procedural form:

```
if(direction=f_to_c) then c:=(f-32)*5/9
'elseif'.:=c*9/5+32
endif
```

"Expert systems" are knowledge-based systems that automate tasks that would normally need human expertise in a certain sector. Expert systems can be divided into two types: those that rely on "deep knowledge" and those that rely solely on "surface knowledge". Humans adhere to all of the established rules of thumb. Professionals are thought to have only a rudimentary understanding of the subject. Typically, such rules are given as "if–then" production rules, and they are highly specific to their particular domains (e.g. diagnosis of a certain group of disorders). Surface information is given in this manner by the "rule-based expert systems" that are growing popular. Deep knowledge, on the other hand, contains a model of a specific universe as well as axioms and laws that can be brought into play

'construct inferences … what rules can do'.

In areas where even human experts' competence is fairly superficial, there appears to be little possibility of constructing anything more than rule-based expert systems (e.g. medical diagnosis). The AI challenge with such deep expert systems is determining which knowledge and models are relevant to a given situation. Because of the properties of knowledge-based systems, their evaluation differs from that of traditional algorithmic software. In one of the few papers that address the problem of quality assurance for AI software (especially expert systems), Green and Keyes discussed the issues as follows: "Expert system software requirements are sometimes nonexistent, erroneous, or rapidly changing." When a user does not fully know his or her own requirements, expert systems are frequently purchased. Some procurements do not include requirements specifications because they are too restrictive or cost-prohibitive. Refinement and consumer contact are used to create expert systems.

'May change quickly or go unnoticed'.

"For verification to succeed, the superior specification's needs must be at least recognised in the sub-species". Ordinate specification: If this isn't the case, tracing requirements is pointless. Expert systems are often created by prototyping and refining a system specification or an informal specification. Intermediate specifications either are not created, are insufficiently precise, or are too changeable to be effective in verification. "Even if comprehensive requirements tracing specifications were available, conventional verification is unlikely to give satisfactory results". There were numerous responses as to whether the implemented system met the requirements. "Traditional validation necessitates meticulous testing methodologies." As long as sufficiently clear requirements and design criteria can be met, the test method preparation should be prioritised. There is no greater difficulty than with traditional software. The test process design becomes a guessing game when requirements and design information are absent, inaccurate or changeable.

"For evaluating the results of expert system tests, there is no commonly accepted, reliable procedure."

'The method of having human experts … a number of problems'.

When independent review is required, there may be no expert accessible or the expert may not be independent.

"Human experts can be biased or narrow-minded. It's possible that the expert system be built to address an issue that no person can solve consistently or efficiently." Issues in evaluating AI software's behaviour all methodologies for estimating software dependability and Part I's the availability of testing on the fly (and, for that matter, mathematical verification) is a document including requirements and specifications, at least to the extent those documents are

'Issues in evaluating AI software's behaviour … those documents are available'.

They may be used to decide whether a program has failed. The problem with AI software requirements and specifications is that there aren't enough of them; therefore, faults in deployed AI systems may go undiscovered since users aren't sure what "proper" behaviour is. Almost any output can appear logical at the time of production, yet subsequently be revealed to be incorrect (for example, during an autopsy or the dismantling of an engine). Dynamic testing for AI software has the same issues: It's not always clear whether the results of a test are sufficient. As a result, before we can apply software dependability, we must handle the challenges of getting software requirements and specifications, as well as dynamic testing, for AI software, as well as evaluating the system against these needs and specifications.

i. **Specifications and Requirements**: The lack of precise needs and specification documents for much AI software reflects the challenge of creating a priori expectations and needs for a system whose capabilities would expand with time. If any of the existing SQA procedures and techniques are to be used in AI software, certain criteria and needs are required. To break free from this deadlock, we recommend separating AI software's "inherently AI" (AI-1) components from the more traditional parts that should be subject to standard SQA.

ii. **Evaluating Desired Competency Requirements**: We've seen how distinguishing the concepts of service needs and minimal competency standards can help to alleviate some of the issues in analysing the behaviour of AI software. For some types of demands, formal or at least detailed, statements of needs and specifications should be achievable and system behaviour may be assessed in respect of these assertions. However, the intended competency criterion could not allow for a precise description, and the only way to assess it might be to compare it to the performance of human specialists.

AI Systems' Acceptance: Even if an AI system outperforms human experts in formal examinations, it's possible that its users will reject it. The previous history of R1, a system that configures the components of VAX machines, is discussed by Gaschnig et al. R1 was expected to generate 50 test orders for a panel of 12 human specialists to review, as part of the acceptance approach. Deficiencies would be corrected, and the cycle would be repeated every 3 weeks with new sets of 50 orders until a satisfactory result. It was possible to reach a high level of precision and dependability. In practice, each review cycle's number of test cases was reduced from 50 to 10, with those ten test cases containing only the most recent ten orders received.

R1 was assessed to be sufficiently skilled to be employed routinely in the conjurations task after five iterations of evaluation (for a total of 50 test cases). Despite the fact that R1 was apparently in use, a human expert was discovered a year later, inspecting and amending 40%–50% of R1's set-ups. It was uncertain whether the VAX computer systems were installed according to R1's precise designs. When questioned, they provided "quite important feedback, albeit a minor one". Little late, on what's important and what's not when it comes to completing the setup work. The following are some of the lessons to be grasped from this experience:

1. The test selection criteria were naive: Only the ten most recent orders were considered. Several test cases were simple, and no effort was made to look for difficult set-up chores that would cause the system to fail. As a result, McDermott admits:

   > In retrospect, it's evident that R1 was still a relatively unskilled configure at the end of the validation step. It had only seen a small portion of the set of possible orders, and as a result, its knowledge was still relatively limited.

2. There was no one-size-fits-all "gold standard". It was discovered that the human evaluators couldn't agree on how to correctly do the set-ups.
3. Testing for development purposes was mixed up with acceptability testing.
4. The eventual users were not sufficiently involved in the system's testing and exercise.

10.8 SOFTWARE SAFETY

Product and system become dependent on software components. 'To create a system … any software components'.

Is It Possible to Fail?: If well-tested software and well-written software can't fail, 'we have believe on this'.

'Experienced based software … to fail actually'.

Hardware does fail, but software does not fail at the same time.

'Hardware failure behaviour … from the world'.

'If software can fail…failure of hardware'.

Software that Is Based on Critical Safety: It's not the same as safety-critical hardware or non-critical software.

Software Failure Modes: Software is a critical application system to tend the fail, 'where expect least'.

Although software does not break, it must be dealt with 'when it does i/p condition' It causes software failures.

Dealing of the task … through the program'.

iii. Anomalous condition I/P is due to the following:
   - Failure of the hardware.
   - Problem of the timing.
   - Unexpected environment condition.
   - Bad user input (I/P).
   - In condition multiple changing.

10.9 CHALLENGES OF THE RESEARCH

'A quite example to admit academic … first high dimension-based problem'.

It will take a small example for practice.

'We considered some problems … a given value'.

'We can control … point extent'. We don't know anything about decision problem for safety relation with SILs. It may choose your favourite classification method as ANN. We assume and provide safety argument acknowledgement according to safety, for example IEC61508. We can also guide and give reason for validating how assumptions may be checked in practice. It has a simple problem and high leverage. 'It does not provide safety parameter … for AI system' If we check certain problem-based condition, it will generalise the same approach for higher dimensions.

10.10 CONCLUSIONS

In this book chapter, we described and gave possible approaches for safety assessment of AI systems. Several questions remain open and are solved as separate applications. A SIL system determines E/E/PE system as normal.

'This hazard is substantiated and risk analysed based'.

It is very compulsory if the requirement of the system is not a SIL system. But in this case, AI can easily be used for this type of situation

'if it has no occurred … a risk analysis'.

It is not compulsory to implement a SIL level system for safety assessment. We have proposed and analysed an approach of this model. It depends on the type of model, which carry more analyses. The assessment of AI model requires an in-depth analysis model for analysis. It means AI can't be analysed and covers lots of different approaches. 'In this case…complicate analysis'. Used in critical systems, it has a restricted approach and types of models to design and simplify artificial intelligence systems. Mackenzie and Pearl (2018) introduced an approach

'the similar angle types of problem … before it's rely'.

The main conclusion is to "formulate a model of the data-generating process, or at least some parts of that process", to show how to provide academic examples in order to proceed the specific types of models.

'We introduce in this book chapter…for safety relation with applications'.

'The formulation of a model of data … for generation process'.

'Without the use of AI system … the possibility is'. There are two possibilities: (i) One is the AI system relevant safety and (ii) the other important safety feature for the E/E/PE system is that it assumes full responsibility for safety.

REFERENCES

Allen Currit, P., M. Dyer, and H. D. Mills (1986). Certifying the reliability of software. *IEEE Transactions on Software Engineering* SE-12(1), 3–11.

Anscombe, F. J. (1973). Graphs in statistical analysis. *American Statistician*, 27(1), 17–21.

Baldwin, G. (1987). Implementation of physical units. *SIGPLAN Notices* 22(8), 45–50.

Ballestar, M.T., Grau-Carles, P. and Sainz, J. (2019). Predicting customer quality in e-commerce social networks: a machine learning approach. *Review of Managerial Science*, 13(3), 589–603.

Baron, S. and C. Feehrer (1985). *An Analysis of the Application of AI to the Development of Intelligent Aids for Light Crew Tasks. Contractor Report 3944*, NASA Langley Research Center, Hampton, VA.

Beam, A.L. and Kohane, I.S. (2018). Big data and machine learning in health care. *JAMA*, 319(13), 1317–1318.

Bellman, K. L. and D. O. Walter. (1987). Testing rule-based expert systems. Course Notes for Analyzing the Reliability and Performance of Expert Systems, UCLA Extension.

Bhavsar, P., Safro, I., Bouaynaya, N., Polikar, R. and Dera, D., (2017). Machine learning in transportation data analytics. In *Data Analytics for Intelligent Transportation Systems* (pp. 283–307). Elsevier.

Boughaci, D. and Alkhawaldeh, A.A. (2020). Appropriate machine learning techniques for credit scoring and bankruptcy prediction in banking and finance: A comparative study. *Risk and Decision Analysis*, 1–10.

Braband, J, H. Gall, H. Schäbe (2018). Proven in use for software: Assigning an SIL based on statistics. In *Handbook of RAMS in Railway systems – Theory and Practice*, Q. Mahboob, E. Zio (Eds.), Boca Raton, FL, Taylor and Francis, Chapter 19, pp. 337–350.

Brunette, E.S., R.C. Flemmer, C.L. Flemmer (2009). A review of artificial Intelligence, *Proceedings of 4th International Conference on Autonomous Robots and Agents, Feb. 10–12, 2009, Wellington*, pp. 385–392.

Char, D.S., Shah, N.H. and Magnus, D. (2018). Implementing machine learning in health care addressing ethical challenges. *The New England Journal of Medicine* 378(11), 981.

Chen, S. H., A. J. Jakeman, J.P. Norton (2008). Artificial Intelligence techniques: An introduction to their use for modelling environmental systems. *Mathematics and Simulation* 78, 379–400.

Corni, M. (2019). *Is Artificial Intelligence Racist? (And Other Concerns)*, 817fa60d75e9, retrieved on October 25, 2018.

Cybenko, G. (1989). Approximations by superpositions of sigmoidal functions. *Mathematics of Control, Signals, and Systems* 2(4), 303–314.

Davis, R. (1980). Teiresias: Applications of meta-level knowledge to the construction, maintenance, and use of large knowledge bases. PhD thesis, Computer Science Department, Stanford University, CA, 1976. Reprinted with revisions in *Knowledge-Based Systems in Artificial Intelligence*, R. Davis and D. B. Lenat, (Eds.), McGraw-Hill, New York, NY.

Edwards, J.R. (2007). Polynomial regression and response surface methodology. *Perspectives on Organizational Fit*, 361–372.

Elmurngi, E. and Gherbi, A. (2017). An empirical study on detecting fake reviews using machine learning techniques. In *2017 Seventh International Conference on Innovative Computing Technology (INTECH)* (pp. 107–114). IEEE.

Ermagun, A. and Levinson, D. (2018). Spatiotemporal traffic forecasting: review and proposed directions. *Transport Reviews* 38(6), 786–814.

Freedman, D.A. (2009). *Statistical Models: Theory and Practice*. Cambridge University Press.

Genesereth, M. R. and N. J. Nilsson. (1987). *Logical Foundations of Artificial Intelligence*. Morgan Kaufmann Publishers Inc., Los Altos, CA.

Goodfellow, I., Bengio, Y. and Courville, A. (2016). *Deep Learning*. MIT Press.

Harmon, P. and D. King (1985). *Expert Systems: Artificial Intelligence in Business*. John Wiley and Sons, New York, NY.

Hättasch, N., N. Geisler (2019). *The Deep Learning Hype, Presentation at 36C3*.

IEC 61508 Functional Safety of Electrical/Electronic/Programmable Electronic Safety-Related Systems, 2010.

Ivanov, A.I., E.N. Kuprianov, S.V. Tureev (2019). Neural network integration of classical statistical tests for processing small samples of biometrics data. *Dependability (Moscow)* 19(2), 22–27.

Kabir, H.D., Khosravi, A., Hosen, M.A. and Nahavandi, S. (2018). Neural network-based uncertainty quantification: A survey of methodologies and applications. *IEEE Access*, 6, 36218–36234.

Kapur, D. and D. R. Musser (1987). Proof by consistency. *Artificial Intelligence*, 31(2), 125–157.

Kou, G., Chao, X., Peng, Y., Alsaadi, F.E. and Herrera-Viedma, E. (2019). Machine learning methods for systemic risk analysis in financial sectors. *Technological and Economic Development of Economy*, 25(5), 716–742.

Leo, M., Sharma, S. and Maddulety, K. (2019). Machine learning in banking risk management: A literature review. *Risks*, 7(1), 29.

Leveson, N. G. (1984). Software safety in computer controlled systems. *IEEE Computer* 17(2), 48–55.

Leveson, N. G. (1986). Software Safety: Why, What and How. *ACM Computing Surveys* 18(2), 125–163.

Leveson, N. G. and P. R. Harvey (1983). Analyzing software safety. *IEEE Transactions on Software Engineering* SE-9(5), 569–579.

Lewis, R.J. (2000). An introduction to classification and regression tree (CART) analysis. In *Annual Meeting of the Society for Academic Emergency Medicine in San Francisco, California* (Vol. 14).

Li, X., Zhang, W. and Ding, Q. (2019). Deep learning-based remaining useful life estimation of bearings using multi-scale feature extraction. *Reliability Engineering & System Safety* 182, 208–218.

Liaw, A. and Wiener, M. (2002). Classification and regression by random forest. *R News* 2(3), 18–22.

Minh, L.Q., Duong, P.L.T. and Lee, M. (2018). Global sensitivity analysis and uncertainty quantification of crude distillation unit using surrogate model based on Gaussian process regression. *Industrial & Engineering Chemistry Research*, 57(14), 5035–5044.

Mosavi, A. and Varkonyi, A. (2017). Learning in robotics. *International Journal of Computer Applications*, 157(1), 8–11.

Murphy, K.P. (2012). *Machine Learning: A Probabilistic Perspective*. MIT Press.

Musa, J. D. A. Iannino, and K. Okumoto (1987). *Software Reliability: Measurement, Prediction, Application*. McGraw Hill, New York, NY.

Panch, T. Szolovits, P. and Atun, R. (2018). Artificial intelligence, machine learning and health systems. *Journal of Global Health*, 8(2).

Pearl, J., D. Mackenzie (2018). *The Book of Why*, Penguin Science.

Polydoros, A.S. and Nalpantidis, L., (2017). Survey of model-based reinforcement learning: Applications on robotics. *Journal of Intelligent & Robotic Systems*, 86(2), 153–173.

Putzer, H. (2019). Ein strukturierter Ansatz für funktional sichere KI, *Presentation at DKE Funktionale Sicherheit, Erfurt*.

Rath, M., (2020). Machine learning and its use in e-commerce and e-business. In *39 Handbook of Research on Applications and Implementations of Machine Learning Techniques* (pp. 111–127). IGI Global.

Reiter, R. (1987). A theory of diagnosis from first principles. *Artificial Intelligence* 32, 57–95.

Schäbe, H. (2018). SIL apportionment and SIL allocation. In *Handbook of RAMS in Railway systems – Theory and Practice*, Q. Mahboob, E. Zio (Eds.), Boca Raton, FL, Taylor and Francis, Chapter 5, pp. 69–78.

Schutte, P. C. and K. H. Abbott (1986). An artificial intelligence approach to onboard fault monitoring and diagnosis for aircraft applications. In *Proceedings, AIAA Guidance and Control Conference, Williamsburg, VA*.

Smola, A.J. and Schölkopf, B. (2004). A tutorial on support vector regression. *Statistics and Computing* 14(3), 199–222.

Stone, P. and Veloso, M. (2000). Multiagent systems: A survey from a machine learning perspective. *Autonomous Robots* 8(3), 345–383.

Underwriter Laboratories: Standard for Safety for the Evaluation of Autonomous Products, draft UL 4600, 2019.

Wang, J. J, Y. Ma, L. Zhang, R. X. Gao, D. Wu (2018). Deep learning for smart manufacturing: Methods and applications. *Journal of Manufacturing Systems* 48, 144–156.

Wigger, P. (2018). Independent safety assessment – Process and methodology. In *Handbook of RAMS in Railway Systems – Theory and Practice*, Q. Mahboob, E. Zio (Eds.), Boca Raton, FL, Taylor and Francis, Chapter 5, pp. 475–485.

Zantalis, F., Koulouras, G., Karabetsos, S. and Kandris, D. (2019). A review of machine learning and IoT in smart transportation. *Future Internet* 11(4), 94.

Zio, E. (2009). Reliability engineering: Old problems and new challenges. *Reliability Engineering & System Safety* 94(2), 125–141.

11 Study and Estimation of Existing Software Quality Models to Predict the Reliability of Component-Based Software

Saurabh Sharma and Harish K. Shakya
Amity University

Ashish Mishra
Gyan Ganga Institute of Technology and Sciences

CONTENTS

11.1 INTRODUCTION

Techniques for analysing the properties of a software design or system are useful for both functional and quality properties (e.g. accuracy, reliability, performance and security).

Predicting the quality properties of a software system using design models can help not only to make the system more trustworthy, but also to save large amounts of money, time and effort by avoiding the implementation of software architectures that do not fulfil the quality criteria. One of the most essential qualities of a software system is reliability, which is defined as the chance of failure-free operation over a

DOI: 10.1201/9780367816414-11

specified time period. A software system's failure tolerance mechanisms (FTMs) are frequently incorporated. They are a key tool for increasing system reliability.

FTMs can be used to mask errors in systems and prevent them from causing failures at various abstraction levels (e.g. source code level with exception handling; architecture level with replication) [2]. Analysing the impact of architectural-level FTMs on component-based software system stability is difficult because:

- FTMs can be used in several aspects of a system's architecture.

    Multiple points in the system architecture can usually be altered to generate architecture variants, such as replacing components with more reliable variants and running components concurrently to increase speed.
- The system's dependability is determined by its design and usage profile (i.e. component services, control flow transitions between them, and sequences of component service calls) [3], in addition to the component's reliability. For example, if faulty code is never executed under a specific usage profile, no errors occur, and users believe the system to be dependable. Existing reliability prediction methodologies for component-based systems do not frequently allow for FTM modelling (e.g. [4–6]) or have limited FTM expressiveness (e.g. [7,8]). These systems lack the flexibility and explicit expression of how FTM error detection and handling affect component control and data flow. An undetected error from a component's provided service, for example, results in no error handling, affecting control and data flow within component services that use this provided service. As a result, when it comes to merging FTMs with the system design and consumption profile, these approaches are constrained. Other approaches (e.g. [9–11]) provide a more extensive study of individual FTMs. These "non-architectural" models, on the other hand, do not represent the system architecture or usage profile. As a result, they are ineffective for determining how individual FTMs used in various portions of a system design affect overall system reliability, particularly when testing for architecture variants under changing usage profiles.

Contribution: This paper offers an explicit and flexible definition of reliability-relevant behavioural aspects (i.e. error detection and error handling) of software FTMs, as well as an efficient evaluation of their reliability impact in the context of the whole system architecture and usage profile, based on the core model (i.e. fundamental modelling steps and basic modelling elements) of our previous work [12]. Our method provides a dependability modelling schema with developer-friendly modelling features (for example, provided/required services, components and connections). For reliability predictions and sensitivity studies, we provide a reliability prediction tool that automates the transformation of models based on the schema into Markov models. In two case studies, we validate our technique and show how it may be used to support design decisions.

The remainder of this chapter is structured as follows: Section 11.2 examines the related work. The phases in our approach are described in Section 11.3. Our dependability modelling schema is detailed in Section 11.4. The transformation used to generate Markov models for dependability predictions is described in Section 11.5. Case studies are used in Section 11.6 to exemplify our methodology. Our assumptions and constraints are discussed in Section 11.7, and the study is concluded in Section 11.8.

11.2 RELATED WORK

Our approach belongs to architecture-based software reliability. Software systems are treated as a collection of software components in modelling and prediction. It has to do with architectural-level fault tolerance modelling and individual FTM reliability modelling.

Several writers have reviewed the area of architecture-based software dependability modelling and prediction [13–15]. Cheung's [3] technique, which uses Markov chains, is one of the first.

Recent work builds on Cheung's work by combining reliability and performance analysis [16] and supporting compositionality [6]; however, it ignores FTMs. Other approaches, such as Cheung et al. [17], which focuses on individual component reliability, Zheng et al. [18], which aims at service-oriented systems, Cortellessa et al. [4] and Goseva et al. [5], which use the UML modelling language, do not address FTMs.

Several approaches in the field explicitly examine error propagation to relax the assumption that a component failure causes a system failure instantly [19–22]. They use error propagation probabilities to simulate the risk of component failures spreading. The sum of these probabilities can be used to describe the chance of component failures being hidden. FTMs, with their error detection and handling, cannot, however, be explicitly considered by these approaches.

Some approaches take a step forward in addressing the issue of including architectural-level FTMs in architecture-based reliability prediction models. Sharma et al. [7] accounted for component restarts and retries in modelling. Different architectural styles, including fault tolerance architectural style, are supported by Wang et al. [8]. These approaches, on the other hand, ignore the effects of FTM error detection and treatment on component control and data flow. Brosch et al. [23] provided a flexible technique to add FTMs; however, they ignored the effects of FTM fault detection on component control and data flow. When the behaviour of FTMs deviates from the precise instances stated by the authors, ignoring the influences of either error detection or error treatment on the control and data flow within components might lead to inaccurate prediction results.

The dependability modelling of individual FTMs has received a lot of attention in the past. Dugan et al. [9] used fault tree approaches and Markov processes to analyse both hardware and software failures for distributed recovery blocks (DRBs), N-version programming (NVP) and N self-checking programming (NSCP). Kanoun et al. [11] used generalised stochastic Petri nets to evaluate recovery blocks and NVP. To evaluate DRB, NVP and NSCP, Gokhale et al. [10] employed simulation rather than analysis. Their so-called non-architectural models aren't accurate representations of the system architecture and usage profile. As a result, while these methods provide a more detailed analysis of individual FTMs, they are limited in their application scope to system fragments rather than the entire system architecture (which is typically made up of multiple structures) and are not suitable for evaluating architecture variants under varying usage profiles.

11.2.1 PRELIMINARY WORK

We published a reliability prediction approach for component-based software systems in [12], which takes into account error propagation for various execution models, such as sequential, parallel and primary-backup fault tolerance executions. However, our fault tolerance modelling support was previously confined to primary backup FTMs, whereas in this work, we can model vast classes of existent FTMs (e.g. exception handling, restart-retry, primary-backup and recovery blocks).

Furthermore, this paper extends the fault tolerance modelling support for multi-version programming FTM supports for modelling composite components and looping structures with discrete probability distributions of loop counts, a more thorough validation and a far more detailed description and discussion of the approach than our previous work [1].

11.3 COMPONENT-BASED RELIABILITY PREDICTION

Component developers and software architects are kept separate in component-based software engineering (CBSE). Component developers create and implement components, as well as give component functional and quality specifications (i.e. models). Software architects can assemble components and test their compatibility using component functional requirements alone. However, component quality requirements must be used by software architects to reason about quality aspects such as dependability, performance and security in component-based software architecture.

Component developers must produce component reliability requirements in our method by detailing how a component's given services are referred to as necessary services in terms of probabilities, frequencies and parameter values. By simply combining these specifications without referring to component internals, software architects can develop a flow and data control model throughout the entire architecture for reliability forecasts.

Our method is depicted in Figure 11.1 as six steps. Component developers build component dependability specifications in the first step. Component developers provide models for components, services and service implementations, as well as failure models for internal operations in service implementations (i.e. distinct failure kinds and their occurrence probability). Different fault tolerance structures (FTSs), such as RetryStructures, MultiTryCatchStructures or MVPStructures (see Section 11.4.2.3), can be incorporated directly into service implementations previously modelled components or as extra components by component developers/software architects. Different configurations are supported by FTSs, such as the number of retries in a RetryStructure, the number of replicated instances in a MultiTryCatchStructure for managing specific failure scenarios and the number of versions executed in parallel in an MVPStructure.

Step 2 involves the creation of a system dependability model by software architects. The system architecture is modelled first, followed by the usage profile. Section

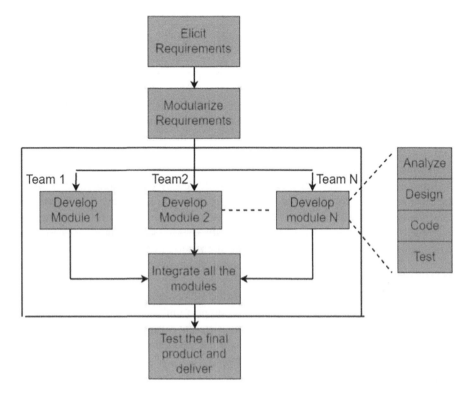

FIGURE 11.1   Component-based software engineering.

11.4 introduces our reliability modelling schema, which aids in the creation of component and system reliability models.

Step 3 involves automatically converting the system reliability model and component reliability specifications into Markov models. By studying the Markov models in Step 4, a reliability prediction and sensitivity analysis may be derived. We provide a reliability prediction tool to help Steps 3 and 4, and the transformation for reliability prediction is discussed in Section 11.5. Sensitivity assessments can also be performed with the tool support, for example, by modifying the reliability-related probability of components within the system architecture to obtain appropriate reliability forecasts. Step 5 is executed if the expected dependability does not meet the reliability criterion. Step 6 is carried out if this is not the case. Step 5 offers several options: component developers can revise components, such as changing FTS configurations; software architects can revise the system architecture and usage profile, such as experimenting with different system architecture configurations, replacing some key components with more reliable variants or appropriately adjusting the usage profile. Sensitivity assessments can be used as a guide for these possibilities, such as identifying the most crucial aspects of the system architecture that need extra attention during revision. Step 6: The modelled system satisfies the dependability requirement, and software architects create the real component implementations using the system architecture model as a guide.

11.4   RELIABILITY MODELLING

11.4.1   BASIC CONCEPTS

An error, according to Avizienis et al. [24], is a portion of the system state that can cause a failure. A defect is the source of the error. When an error causes the delivered service to diverge from the right service, it is called a failure. The deviation might appear in a variety of ways, depending on the sort of malfunction in the system.

The authors outline the principle of FTMs in the same paper. Error detection and system recovery are used to carry out an FTM. Error detection is the process of determining whether or not an error has occurred.

From system recovery, error handling is followed by fault handling. Error management removes errors from the system state, for example, by restoring the system to a previously saved state. Fault handling prevents failures from re-occurring, for example, by replacing failed components with spares or reassigning jobs to non-failed components. Error detection has two main sorts of failures: (i) signalling the presence of an error when none exists, i.e. false alarm; (ii) not signalling the presence of an error, i.e. undetected error

To better model and predict the reliability of component-based systems using architectural-level FTMs, it is necessary to support multiple failure types of a component service and different failure types of different component services, as well as take into account both the influences of error detection and error handling of FTMs on control and data flow within components.

We introduce our reliability modelling schema for characterising reliability-relevant properties of component-based systems in the following section. It would have been possible for us to use UML to structure our approach. However, by incorporating our dependability modelling schema, we avoid the UML's complexity and semantic ambiguities, which make an automated transformation from UML to analysis models difficult. Because our schema is restricted to concepts essential for dependability prediction, it is better suited to our needs than UML enhanced with MARTE-DAM profile [25]. As a result, in the general scenario, our approach can provide an automated transformation for dependability prediction.

11.4.2   COMPONENT RELIABILITY SPECIFICATIONS

11.4.2.1   Components, Services and Service Implementations

Component developers are obliged to give component reliability specifications in our approach. Figure 11.2 depicts an excerpt from our reliability modelling schema, which includes modelling features that assist component developers in developing component reliability specifications. Modelling elements: Component and Service, respectively, are used by component developers to model components and services. A component can be either a primitive component (PrimitiveComponent) or a composite component (CompositeComponent), both of which have nested inner components and are hierarchically constructed. RequiredService and ProvidedService are used to link components to services.

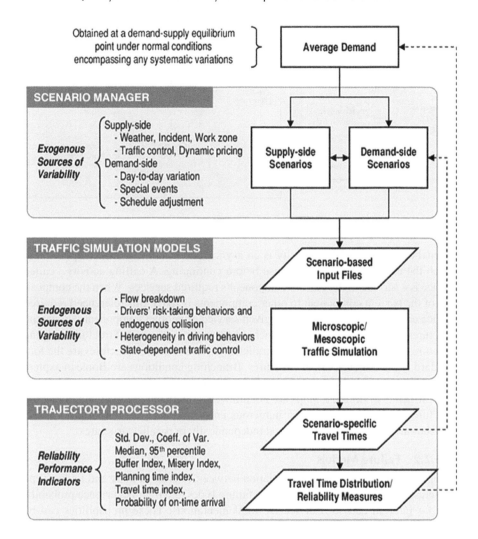

FIGURE 11.2   Modelling elements in our reliability modelling schema.

(i) Exercising: Seven services (from S0 to S6), one composite component (C8), which incorporates three nested primitive components (C5, C6 and C7), and four separated primitive components (from C1 to C4) are shown in Figure 11.3. Component developers must explain the behaviour of each service supplied by a component, i.e. the activities to be performed when a service (Service) in the component's provided services is called, in order to examine reliability. As a result, a component can have several service implementations. Activities (Activity) and structures (Structure) can be part of a service implementation (ServiceImplementation) (Structure). Internal activities (InternalActivity) and calling activities (CallingActivity) are the two categories of activities. The internal calculation of a component is represented by an

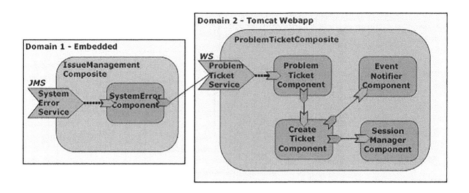

FIGURE 11.3   Example of components and services.

internal activity. A calling activity is an asynchronous call to other components in which the caller waits for a response before continuing. A calling activity's called service is a service in the current component's required services. When the composition of the current component to other components is fixed, this referenced required service can only be replaced by the given service of another component. Sequential structures (SequentialStructure), branching structures (BranchingStructure), looping structures (LoopingStructure) and parallel structures (ParallelStructure) are the four standard types of control flow structures. Branching conditions are Boolean expressions for branching structures. Multiple loops are always limited in looping structures, which is 2; limitless loops are not allowed. Others may be included in looping structures, but they cannot have numerous entry points or be connected. Parallel branches are designed to be executed independently in parallel structures.

11.4.2.2   Failure Models
Component developers use an association between InternalActivity and FailureType to represent failure models (i.e. distinct failure types and their occurrence probabilities) for internal activities of service implementations. These probabilities can be determined using several strategies such as fault injection, statistic testing or growth reliability modelling [13,17].

(ii) Visualisation: Figure 11.4 depicts a service implementation. The service implementation Svc1 includes one internal activity. During the execution of the internal activity, failure type $F2$ can occur with a probability of 0.001617, according to the internal activity failure model.

The service implementation Svc2 has two internal activities (with failure models), four calling activities (to call required services: Svc3, Svc4 and Svc5), one branching structure (with branching conditions: $[Y=\text{true}]$ and $[Y=\text{false}]$) and one looping structure (with loop count: Z).

11.4.2.3   Structures with Fault Tolerance
Detecting errors: To aid in the modelling of FTMs, we have included fault tolerance elements in our FTM modelling schema (FTSs). In FTMs, proper error detection is necessary for proper error handling.

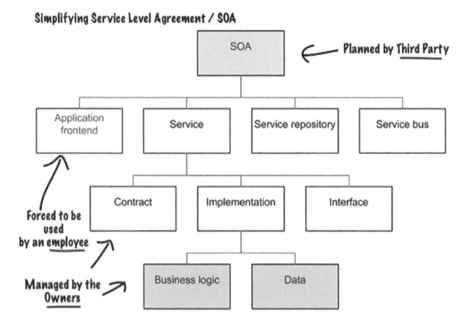

FIGURE 11.4   Example of service implementations.

(iii) Case in point: Figure 11.5 shows an activity with three forms of failure: $F1$, $F2$ and $F3$ (a third failure type, $F0$, is introduced, which corresponds to proper service delivery). Certain failure types, such as $F2$ and $F3$, must be precisely identified in order to provide error handling. For each Fi with i 0, 1, 2, 3, the proportion $ci\,j$ of being recognised as Fj with j 2, 3 must be specified.

As a result, $c0\,j$ is used to symbolise false alarms. False failure signalling is represented by the elements $ci\,j$ and $i\,6=j$. In the case of complete error detection, the error detection matrix has $cj\,j=1$ and $ci\,j=0$ for $i\,6=j$.

RetryStructure: When coping with temporary failures, service re-execution is a good option. A RetryStructure has been created based on this method. There is only one RetryPart in the structure, which contains a variety of activity types, structure types and even nested RetryStructures. The initial execution of the RetryPart simulates regular service execution, whereas subsequent executions simulate service re-execution.

(iv) Visualisation: Figure 11.6 depicts a single RetryPart. During the execution of the RetryPart, failure categories $F1$, $F2$ and $F3$ may occur (the field possible failure types). According to the field handledFailureTypes of this structure, only failure types $F1$ and $F2$ cause the RetryPart to be retried. As many times as the retryCount variable says, the process is repeated (two times in this example).

The concept of a MultiTryCatchStructure stems from the exception handling in object-oriented programming. The structure is made up of two or more MultiTryCatchParts. A single MultiTryCatchPart can contain several activity types, structure types and even nested MultiTryCatchStructures. The first MultiTryCatchPart

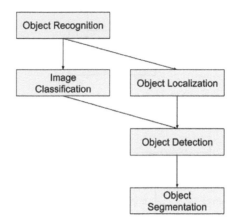

FIGURE 11.5   Error detection semantics for an activity example.

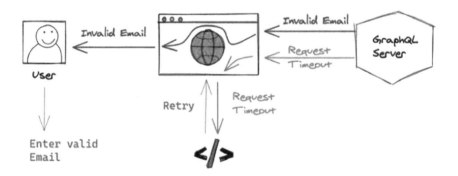

FIGURE 11.6   Semantics for a RetryStructure example.

replicates regular service execution, whereas succeeding MultiTryCatchParts handle specific failure types and do alternate activities, comparable to exception handling catching blocks.

(v) Visualisation: It shows a MultiTryCatchStructure with three MultiTryCatchParts. During the execution of MultiTryCatchPart 1, failure types $F1$, $F2$, $F3$ and $F4$ may occur. Because the field handledFailureTypes of MultiTryCatchPart 2 includes $F2$ and $F3$, and that of MultiTryCatchPart 3 includes $F3$ and $F4$, only failure types identified as $F2$, $F3$ and $F4$ lead to identifying MultiTryCatchParts to handle detected failure types. $F2$ and $F3$ failure types in MultiTryCatchPart 1 lead to MultiTryCatchPart 2, but $F4$ failure type leads to MultiTryCatchPart 3.

During MultiTryCatchPart 2, failure types $F2$ and $F3$ are also available. Furthermore, because the field handledFailureTypes in MultiTryCatchPart 3 includes $F3$ and $F4$, only failure types recognised as $F3$ in MultiTryCatchPart 2 lead to MultiTryCatchPart 3. There is no requirement for an error detection matrix for MultiTryCatchPart 3 because there is no MultiTryCatchPart to manage failures of MultiTryCatchPart 3.

MVPStructure: With a majority voting decision, we designed our MVPStructure utilising the N-version programming (NVP) technique. An MVPStructure contains three or more MVPParts. A single MVPPart can contain many activity types, structure types and even nested MVPStructures. These MVPParts run in parallel in the same environment, similar to how NVP variants (or versions) do: Each accepts the same inputs and produces its own version of the output. The majority voter of the structure then collects the outputs. The system's proper output is assumed to be the majority of the results.

To determine the decision output, the voter must use a collection of results. If the majority of the votes are in disagreement, the voter declares a failure. Otherwise, the agreement creates a result for the voter (i.e. the consensus). The voter's output is right if the majority of the correct outcomes are agreed upon; else, it is erroneous. We assume that the MVPStructure isn't utilised in the same way that NVP isn't used when several separate accurate outputs are possible. The operation of an MVPStructure is depicted.

11.4.2.3.1   Limitations and Assumptions

We presume that components fail independently, as in several similar approaches (e.g. [3,6,16,18]). Without FTMs, a component failure leads to a system failure. This means that the impact of component-to-component error propagation is not taken into account. We refer to our prior work [12] to study the impact of error propagation on reliability prediction of component-based software systems with various execution models, such as sequential, parallel and primary-backup fault tolerance executions.

In our approach, the Markov property of control transitions between components is assumed. This means that the operational and failure behaviours of a component are unaffected by its execution history. Our approach's applicability in many application domains is limited by this Markovian assumption. Many real-world applications, on the other hand, have been proven to satisfy this Markovian assumption at the component level [3]. Our approach can be used to any higher-order Markov model, extending its utility. Because Goseva et al. [14] addressed the issue of Markovian assumption in dependability modelling and prediction in depth, we can validate this. The authors state that a higher-order Markov chain can be mapped into a first-order Markov chain in their paper (i.e. the next execution step depends not only on the previous step, but also on the prior n steps).

Assumptions are made in the evaluation of failure probabilities for internal processes, error detection matrices for FTSs and consumption profiles. There is no such thing as a one-size-fits-all solution to a problem. The bulk of approaches focus on setting up tests to obtain a statistically meaningful amount of data on which to make estimations [26], with component reuse potentially allowing estimations to be based on earlier data. The estimation could be based on the specification and design papers for the system [17]. Estimation could be based on execution traces acquired with profilers and test coverage tools [14] in the last stages of software development, when testing or field data are available.

Our method's argument values are currently fixed constants. They can't be changed in the middle of the game to take into account things such as component condition or system state. Future research will be focused on such considerations.

11.4.2.3.2 Conclusions

In this research, we presented an extended approach for defining reliability-relevant behavioural elements (i.e. error detection and error treatment) of software FTMs and evaluating their reliability impact in the context of the entire system architecture and usage profile.

Software architects use our reliability model to construct a system reliability model, and component developers create component reliability specifications. Then there's a method for forecasting the artefacts' trustworthiness. Two case studies were used to demonstrate the applicability of our technique, emphasising its ability to support design decisions and reuse modelling components for evaluating architecture options under the consumption profile. This form of help can lead to more reliable software systems at a cheaper cost by avoiding potentially considerable expenses for late life cycle upgrades for reliability enhancements.

We plan to combine all of our past work [12], add more intricate error propagation for concurrent executions, add more software FTSs and test our technique more thoroughly. These additions will increase the applicability of our method.

For reliability prediction and sensitivity evaluations, our system automatically converts them to Markov models.

REFERENCES

[1] S. McConnell, Rapid Development, Microsoft Press, 1996.

[2] S. McConnell, Software Project Survival Guide, Microsoft Press, 1997

[3] H. Gumuskaya, Core issues affecting software architecture in enterprise projects, *World Academy of Science, Engineering and Technology*, 9, 32–37, 2005.

[4] Suman and M. Wadhwa, A comparative study of software quality models, *International Journal of Computer Science and Information Technologies*, 5, 4, 5634–5638, 2014.

[5] F. Deissenboeck, E. Juergens, K. Lochmann, and F. Informatik, Software quality models: Purposes, usage scenarios and requirements, In *Proceedings of the ICSE Workshop on Software Quality*, pp. 9–14, 2009, https://doi.org/10.1109/WOSQ.2009.5071551.

[6] M. Yan, X. Xia, X. Zhang, L. Xu, D. Yang and S. Li, Software quality assessment model: A systematic mapping study, *Science China Information. Sciences*, 62, Article no. 191101, 2019. https://doi.org/10.1007/s11432-018-9608-3.

[7] ANSI/IEEE Std 729-1983, *IEEE Standard Glossary of Software Engineering Terminology*, The IEEE, Inc, New York.

[8] ISO/IEC 25010:2011(en) Systems and software engineering — Systems and software Quality Requirements and Evaluation (SQuaRE) — System and software quality models, 2015.

[9] M. Jedlicka, O. Moravcik and P. Schreiber, *Survey to Software Reliability*, Central European Conference on Information and Intelligent Systems, CECIIS, pp. 1–5, 2008.

[10] S. Khatri R.S., Chillar and A. Chhikara, Analyzing the impact of software reliability growth models on object oriented systems during testing, *International Journal of Enterprise Computing and Business Systems*, 2, 1, pp. 1–10, 2012.

[11] J. P. Miguel, D. Mauricio and G. Rodriguez, A review of software quality models for the evaluation of software products, *International Journal of Software Engineering & Applications (IJSEA)*, 5, 6, pp. 31–53, 2014. https://doi.org/10.5121/ijsea.2014.5603.

[12] J. Gao, H.-S. Tsao, and Y. Wu, *Testing and Quality Assurance for Component-Based Software*, Artech House, 2003.

[13] X. Cai, M. R. Lyu, K.-F. Wong, and R. Ko, Component-based software engineering: Technologies, development frameworks, and quality assurance schemes, In *Proceedings of 7th Asia-Pacific Conference on Software Engineering (APSEC)*, pp. 372–379, 2000.

[14] F. Brosch, H. Koziolek, B. Buhnova, and R. Reussner, *Parameterized Reliability Prediction for Component Based Software Architectures*, QoSA 2010, LNCS 6093, pp. 36–51, 2010.

[15] O. Dahiya, K. Solanki, S. Dalal and A. Dhankhar, An exploratory retrospective assessment on the usage of bio-inspired computing algorithms for optimization, *International Journal of Emerging Trends in Engineering Research*, 8, 2, 414–434, 2020. https://doi.org/10.30534/ijeter/2020/29822020.

[16] Pressman, R. S. *Software Engineering a Practitioner's Approach*, 7th Ed. McGraw-Hill, Inc., 2012.

[17] D. P. Narayani, and P. Uniyal, Comparative analysis of software quality models, *International Journal of Computer Science and Management Research*, 2, 3, 1911–1913, 2013.

[18] D. Jamwal, Analysis of quality models for organizations, *International Journal of Latest Trends in Computing*, 1, 2, 19–23, 2010.

[19] R. S. Jamwal, D. Jamwal and D. Padha, Comparative analysis of different software quality models, In *Proceedings of 3rd National Conference, Computing For Nation Development*, Bharati Vidyapeeth's Institute of Computer Applications and Management, pp. 1–5, 2009.

[20] P. Nistala, K. V. Nori and R. Reddy, Software quality models: A systematic mapping study, In *IEEE/ACM International Conference on Software and System Processes (ICSSP)*, pp. 125–134, 2019.

[21] B. Kitchenham and S. Linkman, The SQUID approach to defining a quality model, *Software Quality Journal*, 6, 211–223, 1997.

[22] M. Ortega, Construction of a systematic quality model for evaluating a software product, *Software Quality Journal*, 11, 3, 219–242, 2003.

[23] R. Marinescu and D. Ratiu, Quantifying the quality of object-oriented design: The factor-strategy model, In *11th Working Conference on Reverse Engineering, Delft*, pp. 192–201, 2004.

[24] F. Khomh and Y. G. Gueheneuc, Dequalite: Building design-based software quality models, In *Proceedings of the 15th Conference on Pattern Languages of Programs*, pp. 1–7, 2008, https://doi.org/10.1145/1753196.1753199.

[25] A. Finne, Towards a quality meta-model for information systems, *Software Quality Journal*, 19, 4, 663–688, 2011.

[26] R. S. Kenett, and E. Baker, *Process Improvement and CMMI® for Systems and Software*, CRC Press, 2010.

12 Performance of Multi-Criteria Decision-Making Model in Software Engineering – A Survey

Shweta Singh and Manish Bhardwaj
KIET Group of Institutions, Delhi-NCR

Samad Noeiaghdam
Irkutsk National Research Technical University
South Ural State University

CONTENTS

DOI: 10.1201/9780367816414-12

12.1 INTRODUCTION

Because so many decisions in our modern lives are based on several factors, it is possible to weigh the numerous criteria and receive all of the weights from expert groups [1]. The structure of the problem and the evaluation of multiple criteria are essential. Certain decisions, such as those pertaining to the construction of a nuclear power plant, were made based on several factors.

Some criteria may have an effect on a particular problem, but in order to arrive at the best solution, all the alternatives must share criteria that obviously lead to greater information and better judgements [2]. It's all about figuring out how to structure and solve multi-criteria problems in order to make better decisions and plans. It is the primary goal of this survey to assist decision-makers when faced with a large number of options for resolving a particular issue [3]. There are several situations in which the decision-maker's desire to distinguish between alternatives is necessary.

There are several ways to look at finding a solution. It may be compared to selecting the "most favoured alternative" of a decision-maker from a list of possible choices [4]. Another way to look at the "solution" is to narrow down the options to a few good ones, or to classify them into different preference groups. All "efficient" or "non-dominant" solutions can be found using an extreme interpretation of the problem.

When there are a lot of factors to choose from, the situation becomes more difficult [5]. Without the addition of the relevant information, a unique optimal solution for an MCDM issue can be found. It is common for an optimal solution's notion to be stifled by the non-dominant options. The property of a non-dominant solution is that no alternative solution can be reached without surrendering at least one criterion [6]. Because of this, the decision-maker can readily select a non-dominant answer.

As a matter of fact, the decision-maker could not have done any worse or better in any of the criteria. While there are many non-dominant answers, the decision-maker's final choice is difficult to make because the set is so large. There have been numerous studies on how to find the optimum answer to a problem using a variety of various ways, and each of the MCDM methods has its own uniqueness, as this one on multi-criteria decision-making (MCDM) shows [7]. An acceptable technique of dealing with a problem can be determined by employing MCDM in many applications.

The goal of decision-making (DM) is to find the best possible solution to a problem [8]. Ultimately, it is up to the decision-maker to research the options and choose from a variety of choices in order to get the desired result.

A statistical, quantitative or survey study could be used to find a solution that meets all the requirements while also minimising any potential controversy over the characterisation of the problem. MCDM focuses mostly on decision-making in order to achieve the best possible outcome when there are numerous preferences [9]. The proliferation of options necessitates a reassessment of prioritisation strategies.

The system's complexity rises as more stakeholders are involved in the design process. Multi-attribute decision-making (MADM) and multi-objective decision-making (MODM) are the two basic types of MCDM [10]. The selection of alternatives is made easier by MADM.

It's up to the individual to decide which option is best. To better understand DM's preferences, economists employ the multi-attribute utility theory (MAUT), a branch of the utility theory that focuses on numerous attributes at once. As part of its utility adaptive approach (UTA), regression analysis and linear programming are used [11]. MAUT uses the principle of attribute independence, while UAT uses the principle of variable independence. When two or more criteria are provided, MODM is utilised to generate a continuous set of solutions.

Constraints placed at various intervals define the bulk of MCDM's work. Either manually or mathematically, constraint values can be retrieved [12]. Depending on the intervals, the information retrieved could be either actual or hazy. Data can be retrieved using a modern MCDM approach, which offers the framework for this [13].

In the MCDM process, selecting an aggregate technique is a critical step in reaching a final conclusion. However, recent developments in MCDM have provided a wide range of evaluation theories and assessment methods [14]. There are no set procedures for making decisions. Aggregation methods are used to determine priorities and rank alternatives depending on the desired and the target of comparison.

12.2 PREVIOUS WORK

Fuzzy logic allows decisions to be made with approximated values despite the lack of complete information. Even if a decision turns out to be bad, it can be changed if further information comes to light later on [15]. There is no way to make a judgement based on logic when there is no information at all. Typical non-fuzzy approaches (e.g. linearisation of nonlinear situations) typically rely on mathematical approximations, which results in poor performance and high costs [16].

Fuzzy systems often outperform traditional MCDM methods in certain situations. A great deal of work has been done in a wide range of areas such as banking and general purpose, student and teacher performances, water resource location and many more of these sectors [17]. A study of the available options has been carried out in order to determine which alternatives are optimal. The explicit consideration of multiple criteria in MCDM structures difficult issues, allowing for better and more well-informed decisions.

12.2.1 Different Approaches of MCDM

To select the optimal alternative, MCDM approaches have been used in a variety of contexts. The method of MCDM and its various forms are shown in a hierarchical structure in Figure 12.1. The following sections provide an overview of the most common MCDM approaches.

12.2.1.1 Analytic Hierarchy Process

AHP is based on the premise that experts' knowledge in a subject may be gathered. An alternate selection and justification problem is approached using the principles of fuzzy set theory and hierarchical structure analysis [18]. Interval judgements are more reliable than fixed value assessments for decision-makers. This strategy can be used when a user choice isn't clearly specified because of its fuzzy nature.

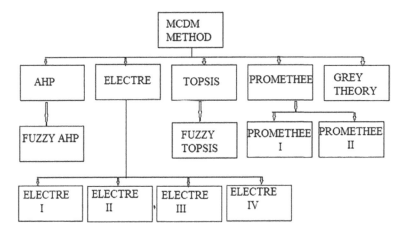

FIGURE 12.1   Basic block diagram of MCDM methods.

It is impossible for AHP to portray human imprecise thoughts because it includes expert judgements and multi-criteria evaluation [19]. The fuzzy set theory makes the comparison process more flexible and capable of explaining the preferences of experts than the standard AHP, which emphasises the clear judgements of decision-makers. Using the AHP, an MCDM problem is broken down into a systematic hierarchy procedure [20]. The structure of an $m*n$ matrix (where m is the number of alternatives and n is the number of criteria) is the focus of the AHP method's last phase.

A matrix is generated based on the relative relevance of each condition. On the basis of priority theory, AHP constructs its hierarchies. Problems involving the simultaneous consideration of multiple criteria or alternatives are dealt with in this book.

12.2.1.2   Fuzzy Analytic Hierarchy Process

It is used in conventional market surveys, etc., to fuzzify analytic hierarchy process (fuzzy AHP). Pairwise comparisons are used to determine the weights of each item's evaluation in AHP, as well as its value in relation to other products and alternatives; however, the results of these comparisons are not 0, 1, but rather a numerical number. If a weight is expressed in fuzzy AHP, then the traditional constraint that the sum of several weights 1 can be loosened is also relaxed [21].

12.2.1.3   TOPSIS

It is assumed by the TOPSIS technique that each criterion has a tendency to monotonically increase or decrease utility, leading to the easy definition of the positive and negative ideal solutions [22].

In order to determine how close, the alternatives are in comparison with the perfect solution and a strategy based on Euclidean distances is put forth. When comparing the relative distances of the alternatives, this will determine which one is preferred.

ELECTRE's non-dimensional criterion is first converted into a non-dimensional criterion through the TOPSIS approach [23]. As outlined in TOPSIS, the chosen alternative should be the lowest distance from PIS (positive) and the greatest distance from NIS (negative). The MCDM is made easier by using this approach of ranking. The criteria in each region are evaluated using the fuzzy TOPSIS approach, and then the criteria are sorted according to the region [24].

12.2.1.4 ELECTRE

Elimination Et Choix Traduisant la REalite is one of the MCDM approaches, and this method allows decision-makers to select the best choice with the greatest advantage and the least conflict in the function of several criteria.

To distinguish between a given set of options, an individual can utilise the ELECTRE method, which was originally known as ELECTRE I. ELECTRE I, II, III, IV and TRI are only some of the variants of ELECTRE that exist. Fundamentally, each approach is founded on the same ideas, but they differ in terms of how they are used and the type of choice problem they are meant to solve [25].

ELECTRE I; ELECTRE TRI; and ELECTRE II, III and IV are all geared towards solving selection; assignment; and ranking problems, respectively. "Outranking relations" are an important concept to remember. Coordination indices can be used to model a decision-making process in ELECTRE.

The concordance and discordance matrices are used to calculate these indices [26]. A concordance and discordance index is used to assess outranking relations among distinct alternatives and to determine the best option based on the clear data [27].

12.2.1.5 Grey Theory

The terms "insufficient data" and "poor knowledge" are both used to describe grey theory, which is a rigorous mathematical examination of systems that are both known and unknown.

There are a large number of input facts that are separate and insufficient when grey theory studies the interactional analysis since the decision-making process is not evident [28]. In recent years, a number of decision-making issues have been solved using the grey theory technique.

It has been commonly used to discover the optimum solution when the number of options and criteria are considerable [29]. The decision-making process dictated the selection of these techniques. It was decided that ELECTRE would be used to choose the best candidates, that TOPSIS would be used to rank them, and that the grey hypothesis would be used to select the best candidate when complete data were not available. Fuzzy MCDM approaches are put to use in the following section [30]. Other MCDM approaches are available, and we've listed some of them below, along with their intended use, pros and cons and other relevant information.

12.2.1.6 ANP

Since its inception in 1996, the ANP technique has allowed the criteria to be interdependent. Most problems cannot be arranged in a hierarchical way because of the contributions from several levels [31]. With the cycles intertwined within the system, ANP is depicted as a network.

Because of the inherent ambiguity in human judgement, ANP is unable to adequately assess essential criteria. Fuzzy ANP uses the fuzzy preference programming method to derive local weights [32]. Weights based on local factors are combined to construct a supermatrix to rank alternatives.

12.2.1.7 VIKOR

To resolve choice difficulties involving several criteria, Serafim Opricovic created VIKOR in 1979 and published an application under the name VIKOR in 1980. A similar ideal solution to TOPSIS is used in the method's aggregation and decision representation [33].

The linear normalising method is employed in the VIKOR software. For the majority, it provides the most usefulness, while allowing each individual to have a small amount of preference over the rest of the population.

12.2.1.8 PROMETHEE

The compound PROMETHEE and its antonym is presenting the correct answer. Rather than presenting the correct answer, geometric analysis for interactive aid (GAIA) established in the 1980s is used to conquer alternative optimal solutions to achieve goals [34]. These techniques aid programmers in the creation of a process structure, analysis of the problem and prioritisation of potential solutions.

12.2.1.9 SMARTER

MAUT-based SMARTER (simple multi-attribute rating technique exploiting ranks) method is mostly used for preference analysis.

SMART (simple multi-attribute rating technique) is a family of compensation approaches established by Edwards and Barron. Ranking criteria are numerically weighted by using the rank order centroid (ROC) in SMARTER [35]. Goal and decision-makers, criterion setting, defining goal alternative and competitors, problem recognition, investigation of noteworthy alternatives and the calculation of the one-dimensional value function are all part of SMARTER (Table 12.1).

12.2.1.10 Wiegers

In a recent work, requirement prioritisation was accomplished using Wiegers method of fuzzy logic. Benefits, penalties, risks and costs all play a role in determining the best strategy [36].

Using the membership function, weights are rated out of five. In order to implement the fuzzy logic, MATLAB® membership functions and designer inference rules are used [37]. Real-time implementation makes it a better fit because of the high priority placed on requirements during the development stage.

Since stakeholders' decisions and the requirement are both unclear and ambiguous, advancement in MCDM approaches suggests that a fuzzy version of the methods is more suited. Research suggests that fuzzy concepts can better handle ambiguity in complex decision-making [38].

12.2.1.11 Previous Research Work

Based on stakeholder-defined independent criteria, Jusoh et al. [39] implemented the AHP for the selection of open-source software (OSS). It is not uncommon for the contributors of different organisations to use different methods of selecting new

TABLE 12.1

Advantage and Disadvantage of MCDM Methods

S. No	MCDM Methods	Description	Disadvantage	Advantage
1.	Weighted product model	Products are compared with each other in reference of weight.	1. For equal weight of DM's no solution	1. Remove any unit of measurement
2.	Analytic hierarchy process	Pairwise comparisons with different criteria	1. More comparisons are required pairwise 2. Ranking irregularities	1. Non-biased decision-making 2. Each element importance becomes clear
3.	Weighted sum model	Same unit alternatives are evaluating	1. Multi-dimensional evaluation is difficult	1. Strong in single-dimensional problems
4.	ELECTRE	Best choice with maximum advantage	1. Consumption of time is more	1. Used outranking
5.	Grey analysis	Deal with incomplete data	1. No optimal solution information	1. Unique solution for perfect
6.	Analytic network process (ANP)	ANP use different alternatives for best solution	1. Uncertain 2. Time-consuming	1. No independency required 2. Accurate prediction
7.	Data envelopment analysis (DEA)	Used to find the combined efforts efficiency	1. No absolute efficiency 2. Demanding large problems	1. Handles multiple input output 2. Comparisons are directly against peers

members. Choosing the right software to address a certain problem is a personal choice for each operator.

Data and service quality are also examined as part of this investigation. The author listed 12 criteria for selection, such as reliability, usability, performance efficiency, functionality and competence. In order to meet the specifications of OSS, the system defined the features. In order to select the OSS, the AHP was used to determine the best option.

Fuzzy theory may be used in the future to express the weights associated with needs in a hierarchical structure. In the future, decision-making by consensus could be utilised to include all relevant parties. AHP and TOPSIS were also mentioned by Vinay Selat [40]. AHP and TOPSIS were used to make decisions based on the outcomes of the integration of goals after prioritisation and evaluation.

Requirements engineering used this proposed methodology to validate various judgements when several stakeholders were engaged. The most important part of the project was the development of frameworks for decision support systems. An e-commerce application was used to demonstrate the proposed method. Future study should take into account various stakeholders, prioritise requirements or hard goals and investigate game theoretic approaches in the decision support system.

In their work on software quality model selection, Sumeet Kaur Sehra et al. [41] highlighted some of the applicability of FAHP. Finding a web development platform, evaluating the quality of a website and determining the success factors of an online store may all be accomplished using the FAHP in this study.

The study evaluated the McCall, Boehm and ISO9126 quality management models using three separate criteria: dependability, efficiency and maintainability. Normalised weights are used to narrow down the pool of potential candidates for the model. Using both the FAHP and AHP, the weights of the criteria are calculated and compared to each other.

For ISO9126, the best software model has a normalised weight of 0.38 for FAHP and a weight factor of 1.39 for AHP, indicating Boehm's model selection. In value assignment, the outcomes are influenced by both the specific application and the decision-maker's point of view. The FAHP approach can be considered one of the greatest solutions for ranking and assessment concerns in software engineering because the decision-making is unclear.

For prioritising requirements, Sahaaya et al. [42] used the ELECTRE approach. It is common practice to use ELECTRE to determine the relative importance of several projects. The 100-point technique and ELECTRE were used to rank the contributions from various stakeholders in this system.

Because of its lower implementation costs and man-hour requirements, the resulting system was found to have an advantage over more traditional systems. The system's flaw is the 100-point technique, which is limited when dealing with huge amounts of criteria. Stakeholders should be taken into consideration when utilising fuzzy approaches in future research.

For the assessment of agile methodologies for small and medium organisations to fulfil the need for software development, Silva et al. [43] introduced a multi-criteria method called SMARTER. In the selection process, DSDM (dynamic systems development method), SCRUM, XP2 and Crystal were considered among the most prominent agile models.

These approaches are the only options available. After defining a set of criteria, a survey was carried out. The final results were obtained by converting the language values into numerical indexes. The multi-attribute values were used to rank the methodology. There is a lack of complete information for the robust selection of the process through this method, which is more time-consuming and expensive.

This is one of the most current studies in the field of software engineering using the SMARTER program. The study has come to an end, and the researcher has made some observations that merit further investigation. For more exact criteria, further research suggests numerical scaling may yield better results than survey techniques.

Future studies should focus on a more efficient quantitative analysis of linguistic scales to evaluate the alternatives. There are significant differences between FTOPSIS and TOPSIS, as described by Elissa Nadia Madi et al. [44] in their discussion of these two approaches. The FTOPSIS method's difficulties and challenges are also discussed in this paper.

As a result of spotting these flaws, a workaround has been proposed that could be implemented in the future to make the interesting fuzzy TOPSIS approaches more reliable. Prioritising interdependent requirements using ANP was proposed by Javed Ali Khan and colleagues [45]. The consistent results that ANP provides, which are proportion scale dependent, lead the researchers to believe that ANP is an excellent tool for determining the priority of a set of requirements.

According to the study, ANP prioritises better than AHP. MATLAB® was used to run the simulation. In the future, it is proposed that ANP be used in the software business to prioritise requirements.

Using a questionnaire method to gather data, Romulo Santos et al. [46] applied the hybrid cumulative voting (HCV) prioritisation strategy. It was decided to focus on a case study of COTS software requirements prioritising. Some of the potential software user's responses were recorded online. Ratio scale weights are calculated from the database using the HCV approach.

MACBETH (measuring attractiveness by the categorical-based evaluation technique) was used to consolidate the results. The method is determined to be able to meet the features of market-driven software development. A case study favouring a worldwide perspective, with the culture and economic weight of the region as additional aspects, is being considered for future work.

Integer linear programming with extra selection criteria such as cost and requirement interdependency is another upgrade that has been presented. Statistics was utilised by Hadeel E. Elsherbeiny et al. [47] to prioritise the needs of a system with many stakeholders. Because the respondents gave it the highest rating of the three techniques of eliciting requirements, the researcher decided to employ Rate P as a means of gathering the information from them.

When using Rate P, the rating scale ranges from 0 to 5, with -1 denoting the absence of a requirement. Group brainstorming and brainstorming sessions are among the methods for obtaining the necessary information. There are 76 participants in the study, 10 project goals, 48 requirements and 104 requirements in detail. There is an input of non-prioritised requirements to the system, and the output is a recommendation for prioritised ones. SPSS is used by the researcher to prioritise and identify correlations to predict the needs of the stakeholders.

Game-based needs prioritisation techniques in software engineering are discussed by Kifetew Meshesha Fitsum et al. Requirements engineers can benefit from the use of the decision-making game (DMGame). It uses gamification and automated reasoning to prioritise requirements and engage stakeholders in the process of making decisions. Automated prediction algorithms are used to make decisions in DMGame, which relies on an online role-playing game (ORPG).

When taking into account the contributions of various stakeholders and automating the prioritisation of tasks, the process was shown to be more efficient. Pairwise comparison is used to rank alternatives in the AHP method for automated reasoning. It's built to manage a variety of people and groups.

An alternative to AHP for many requirements may be a non-pairwise technique using multi-objective optimisation in future work. Various strategies for determining the priority of requirements have been examined by Raneem Qaddoura et al. [48] in depth. Methods are chosen based on the nature of the project and the requirements that must be met. Some of the factors used in the comparison of these methods include the difficulty of use, reliability of outcomes and fault tolerance. The interesting technology will be compared to other data mining and machine learning methods in the future.

Researchers such as Hassan Abeer and Ramadan Nagy [49] described the various ways they've used to rank the importance of system needs. Fuzzy Wiegers' method is used in this study to develop a framework for ranking needs by weighting benefits, penalties, costs and risks. The classical Wiegers' method is compared to a numerical example utilising MATLAB® and a spreadsheet in this study.

Hassan Abeer and Ramadan Nagy [50] proposed a hybrid model for demand prioritising employing three different strategies such as QFD (quality function deployment), CV (cumulative voting) and AHP using fuzzy technique. Due to stakeholder decisions being ambiguous, the concept of employing a fuzzy method was born. Given the ambiguity of decision-making in fuzzy, the real world appears to be a closed-off environment.

For prioritisation purposes, requirements are categorised as large, medium and tiny. When it comes to making complex decisions, this strategy is able to handle group decision-making, as well as the uncertainty that can arise during group decision-making. To ensure that this method can be easily implemented, and efficiently and effectively manage uncertainty in decision-making, author compares the proposed fuzzy version of this method to the classical form.

12.2.2   FMCDM APPLICATION

It is utilised in a wide range of industries, including banking, performance improvement, decision-making in diverse organisations, safety evaluation and multi-choice general-purpose problems. A variety of FMCDM approaches and applications are explored in this section.

12.2.2.1   Fuzzy MCDM Applications

It's common for businesses to become hazy when there are a lot of options accessible to make the greatest decision. When it comes to an organisation's supplier selection, for example, MCDM involves both numerical and qualitative elements.

To find the finest supplier, you must first determine your needs and trade-offs between these observable and ethereal elements, some of which may be in opposition to one another.

An effective supply chain relies heavily on the selection of suppliers who are capable of delivering the proper quality product or service at an appropriate price, at the right time and in an appropriate quantity to consumers. FMCDM approaches such as TOPSIS, ELECTRE and AHP have been used to overcome this problem.

ELECTRE is a tool for moving away from bad points and getting closer to the positive. Marine engineering is fundamentally concerned with safety issues. When it comes to safety in maritime engineering, crew members' understanding of risk and their ability to handle it is critical. Fuzzy approaches such as TOPSIS, ELECTRE and AHP have been used to discover the optimal safety measures. Location planning, resolving issues with OWA operators and other topics covered in Table 12.2 have all been tackled using fuzzy MCDM approaches.

TABLE 12.2
Different Approaches of Fuzzy MCDM

Application	Criteria	Problem	Techniques	Alternatives	Best Alternatives
Location planning for urban distribution centres	1. Cost 2. Security 3. Environmental impact 4. Resource availability 5. Connectivity	For distribution at the urban centres location planning; congestion control is the main issue	TOPSIS	Area 1 Area 2 Area 3	A1>A3>A2 Area 1 is the best
Revising the OWA operator problem	1. Benefit/cost 2. Job creation	Finding the best robust out of available	FSROWA FMCDS	Germi chai Bijar Talvar Givi Klaghan	Germi chai is the best

Enhancing information delivery	3. Allocation of water
	4. Public participation
	  1. Partner's Price range
	  2. Partner's product range
		To find the best supplier for mould and die		P1, P2, P3, P4	P2 is the best

Evaluation of supplier in supply chain management	1. Reliability 2. Product price 3. Ordering cost 4. On-time delivery 5. Financial stability	To find out the correct supplier	TOPSIS	S1, S2, S3, S4	S3 (S3=Supplier 3)
Environmental risk assessment and health safety	1. Technological risk 2. Health safety risk	With AHP find out the best location for power plant	AHP	L1, L2, L3, L4	L3
Assessment of healthcare waste using fuzzy	1. Technical 2. Social 3. Environmental	Evaluation of HCW treatment alternatives using fuzzy with multi-criteria decision-making	Fuzzy MCDM	Microwave Landfill Incineration	Landfill
Multi-criteria decision-making approach based on immune algorithms	Waist girth (W) Hip girth (H)	Solve the garment matching problem	Co-evolutionary immune algorithm for the MCDM model	Seventy shirts of same colour and style were used for study	Which satisfies the user

TABLE 12.3

Performance Evaluation of Fuzzy MCDM

Application	Criteria	Problem	Techniques	Alternatives	Best Alternative
Training performance evaluation of administration science instructors by fuzzy MCDM approach	1. Level of knowledge 2. Way of teaching 3. Individual features	Find out the best trainee	FMCDM	I1, I2, I3, I4	I1
Profitability and customer satisfaction using MCDM	Power quality, reliability, cost, availability	To find out the best tool (MCDM) for investigation for achieving the post	AHP	A1, A2, A3, A4	A2
Teachers performance evaluation and appraisal using MCDM	1. Analysis of growth 2. Impact of environment 3. Analysis of risk	Find out the best company for investment	CPRAS-G	T1, T2, T3, T4	T3

12.2.2.2   Fuzzy MCDM in Performance Evaluation

Although the methodologies are widely used in a wide range of fields, they can also be employed to analyse organisational effectiveness. As shown in Table 12.3, the effectiveness of organisations can be assessed using FMCDM methodologies. The effectiveness of a teacher can be measured using the COPRAS-G approach. It uses a numerical scoring system in the form of interval marking.

Quantitative numerical scores can be handled by common approaches documented in previous research. The COPRAS-G technique, on the other hand, is able to account for interval making given to a specific item.

Fuzzy set theory is used to measure the performance of administrative instructors in the evaluation of their training performance. AHP is used to calculate the weight of the criteria, while TOPSIS is used to rank the results. The fuzzy MCDM approach is used to analyse choice alternatives, including subjective judgements by a group of decision-makers. Individual decision-makers can utilise a pairwise comparison process or a linguistic grading system to help them form comparable judgements.

A performance evaluation is a tool used to gauge an employee's overall contribution to the company. If an employee does or does not satisfy particular criteria, an evaluation might be used to make recommendations for next steps. Uncertainty arises when evaluating performance; hence, the MCDM technique is used to gauge any performance problems. COPRAS-G is used to identify the top teachers in the evolution of teacher performance utilising a variety of criteria and alternatives.

Such factors as knowledge level, problem-solving capabilities and cognitive capacities have been used to evaluate the performance of the training administrative teacher.

Electrical energy is in high demand from consumers due to its ubiquitous presence in human endeavour. Technical and organisational metrics are typically used in the planning and operation phases of electrical power systems to examine appropriate tools (MCDM methods) that enable decision-makers in achieving the goals such as customer satisfaction and profit making. Using interval-valued intuitionist fuzzy sets, a MCDM strategy is utilised to determine the best company to invest money in order to earn greater profit.

12.3 SURVEY RESEARCH OUTPUTS

12.3.1 COMPARISON OF AHP AND FUZZY AHP

12.3.1.1 Analytic Hierarchy

Using AHP, a decision-maker can use many criteria to rank options and select the best one. Using this strategy, decision-makers can narrow their options down to a single, better choice by comparing how well each one fits a minimum set of criteria. Humans aren't very good at making quantitative predictions, but they are equally good at creating quantitative forecasts, thanks to fuzzy AHP. During the process of making decisions, there is a rise in inconsistency between the possible outcomes.

If any of the criteria has a lower importance than the rest, it can be weighed as zero in a fuzzy pairwise comparison, unlike other techniques. The decision-making process does consider that criterion, but it isn't given much weight because there are so many others. To be sure, the classic AHP technique does not allow for the "zero-weighed" condition, but the numerical weight of a criterion will be close to zero if it is judged as being smaller than all the others.

Fuzzy AHP can simply overlook the less important criteria, whereas classic AHP places so much emphasis on them. Fuzzy ARP displaying additional information may benefit from this as well because there is no difference between a criterion's presence and nonexistence in the minds of the decision-makers.

As a result, the decision-maker will be able to focus on more critical factors. Fuzzy approaches and classical algorithms aren't rivals when used under the same conditions. As a general rule, the classical technique should be used if information or evaluations are known to be accurate; if the information or evaluations are not known, the fuzzy method should be used.

In recent years, because of the uniqueness of information and decision-makers, it has been necessary to incorporate the possibility of deviation into decision-making procedures, and as a result, a fuzzy version has been produced for each decision-making approach. This necessity led to the development of the fuzzy AHP approach.

A questionnaire is used to assess a subject's linguistic and affective abilities. Scaled numerical values for each language feature are predetermined. Although these numbers are exact numbers in classical AHP, the fuzzy AHP method uses intervals between two numbers to represent them.

12.4   RESEARCH DIRECTIONS IN MCDM

The actual purpose of an integrated decision-making system is to enable the decision-maker to look into the future and make the best possible decision based on previous and current facts and future projections. Predicting the risk and vulnerability of individuals and infrastructure to both natural and man-induced hazards is an important part of sustainable development.

This necessitates the transformation of data into knowledge and a thorough examination of the outcomes of information consumption, decision-making and participatory procedures. Using fuzzy logic will only provide an approximation of a solution, according to the findings of the research.

Data can be analysed using fuzzy logic for any application, whether it's quantitative or qualitative data. It is possible to carry out a large number of smaller activities by utilising the various FMCDM approaches.

The originality of each strategy is evident. Analysing a software application can be somewhat nebulous in this manner. There have been past attempts to map out what information is needed by different groups of people, such as government agencies needing a lot more data than, say, a customer service department or a corporate management team.

Customers, government and management all have different needs for information, so it's critical that the right information is delivered to them in the format they prefer. Each of these groups may have their own ideas about how information should be delivered; banks can gather the data they need by interviewing a wide range of customers and having them complete various applications and questionnaires.

Uncertainty in user information distribution is now in place. Each user's information is unique, and the information's substance is unique as well. It is imperative that the right information be sent to the right person at the right time via a channel they choose. The level of information and the level of security also change depending on the needs of different users. FMCDM approaches, which are used to deliver the correct information to the right person at the right time, can be utilised to overcome this uncertainty problem.

12.5   CONCLUSIONS

This study identifies potential in MCDM, where numerous choices are involved. Many applications, such as financial, summative assessment, prevention of injury and other multi-criteria domains, make use of fuzzy MCDM. Using FMCDM, we can assess a large number of options using a variety of criteria before settling on the optimal one. For each problem, the MCDM approaches were chosen in accordance. The use of MCDM has only been implemented in a few cases. This survey is focused on the banking industry because of the high level of ambiguity in the decision-making process. MCDM on a fuzzy basis is well suited to problems with approximate solution spaces. A solution can be found by analysing both quantitative and qualitative data in any application using FMCDM. Many methods under MCDM exist, each of which has a distinct set of capabilities, and the method must thus be selected for each specific task.

REFERENCES

[1] Albayrak, E., Erensal, Y. C. (2005). A study bank selection decision in Turkey using the extended fuzzy AHP method. *Proceeding of 35th International Conference on Computers and Industrial Engineering, Istanbul, Turkey.*

[2] Aldlaigan, A., Buttle, F.A. (2002). A new measure of bank service quality. *International Journal of Service Industry Management*, 13, 38–362.

[3] Business Credits (2006). Non-financial data can predict future profitability. *BusinessCredits*, 108(4), 57.

[4] Nikoomaram, H., M. Mohammadi, M. JavadTaghipouria and Y. Taghipourian (2009). *Training Performance Evaluation of Administration Sciences Instructors by Fuzzy MCDM Approach.* Tehran, Iran.

[5] Y. Tansellç (2012). Development of a credit limit allocation model for banks using an integrated Fuzzy TOPSIS and linear programming. Expert System with Applications, 39(5), 5309–5316.

[6] T. Ozcan, N. Celebi (2011). Comparative analysis of multi-criteria decision making methodologies and implementation of a warehouse location selection problem. Expert System with Applications, 38, 9773–9779.

[7] Mohammad SaeedZaeri, Amir Sadeghi, Amir Naderi, et al. (2011). Application of multi criteria decision making technique to evaluation suppliers in supply chain management, *African Journal of Mathematics and Computer Science Research*, 4(3), 100–106.

[8] Schinas O. (2007). Examining the use and application of multi - criteria decision making techniques in safety assessment, *International Symposium on Maritime Safety, Security & Environmental Protection, Athens.*

[9] Anjali Awasthia, S.S. Chauhanb, S.K. Goyalb (2000). *A Multi- Criteria Decision Making Approach for Location Planning for Urban Distribution Centers under Uncertainty.* CIISE, Montreal, Canada.

[10] T.C. Chu (2002). Facility location selection using fuzzy TOPSIS under group decisions, *International Journal of Uncertainty, Fuzziness and Knowledge-Based Systems*, 10(6), 687–701.

[11] Mahdi Zarghami, Ferenc Szidarovszky (2011). *Revising the OWA Operator for Multi Criteria Decision Making Problems under Uncertainty a Faculty of Civil Engineering,* Tabriz, Iran.

[12] DoraidDalalah, Mohammed Hayajneh, Farhan Batieha (2011). A fuzzy multi-criteria decision making model for supplier selection, *Expert Systems with Applications*, 38, 8384–8391.

[13] Yong-Sheng Ding, Zhi-HuaHu, Wen-Bin Zhang (2011). Multi-criteria decision making approach based on immune co-evolutionary algorithm with application to garment matching problem, *Expert Systems with Applications*, 38, 10377–10383.

[14] Y. Peng, Y. Zhang, Y. Tang, S. Li (2011). An incident information management framework based on data integration, datamining, and multi-criteria decision making, *Decision Support Systems*, 51, 316–327.

[15] M. Dursun, E. ErtugrulKarsak, MelisAlmulaKaradayi (2011). Assessment of healthcare waste treatment alternatives using fuzzy multi-criteria decision making approaches, *Expert Systems with Applications*, 38, 10377–10383.

[16] S. Rezaiana, S. Ali Joziba (2012). Health-safety and environmental risk assessment of refineries using of multi criteria decision making method, *APCBEE Procedia*, 3, 235–238.

[17] K. Sadeghzadeh, M. Bagher Salehi (2011). Mathematical analysis of fuel cell strategic technologies development solutions in the automotive industry by the TOPSIS multi-criteria decision making method, *International Journal of Hydrogen Energy*, 36, 13272–13280.

[18] Hung-Yi Wua, Gwo-Hshiung Tzenga, Yi-Hsuan Chen (2011). *A Fuzzy MCDM Approach for Evaluating Banking Performance based on Balanced Scorecard, Taiwan.*

[19] Ne lu, CengizKahramanb (2007). Fuzzy performance evaluation in Turkish banking sector using analytic hierarchy process and TOPSIS. *Expert Systems with Applications*, 36, 11699–11709.

[20] Ashton, C. (1998). Balanced scorecard benefits Nat west bank. *International Journal of Retail and Distribution Management*, 26(10), 400–407.

[21] Tsai, W. H., Yang, C. C., Leu, J. D., Lee, Y. F., & Yang, C. H. (2001). An integrated group decision making support model for corporate financing decisions. *Group Decision and Negotiation*, 1–25. C.T. Chen, A fuzzy approach to select the location of the distribution center, *Fuzzy Sets and Systems*, 118, 65–73.

[22] S.Y. Chou, Y.H. Chang, C.Y. Shen (2008). A fuzzy simple additive weighting system under group decision making for facility location selection with objective/subjective attributes, *European Journal of Operational Research*, 189(1), 132–145.

[23] A. Mazumdar (2010). Application of multi-criteria decision Making (MCDM) approaches on teachers Performance evaluation and appraisal.

[24] Chen, C.W.E. (2000). Extensions of the TOPSIS for group decision-making under fuzzy environment. *Fuzzy Sets and Systems*, 114, 1–9.

[25] Rabah Medjoudj, Djamil Aissan, Klaus Dieter Haim (2013). Power customer satisfaction and profitability analysis using multi-criteria decision making methods, *Electrical Power and Energy Systems*, 45, 331–339.

[26] V. Lakshmana Gomathi Nayagam, S. Murali Krishnan, Geetha Sivaraman, (2011). Multi-criteria decision-making method based on interval-valued intuitionistic fuzzy sets, *The Journal of Analysis*, 1464–1467.

[27] H.C.W. Lau, C.W.Y. Wong, P.K.H. Lau, K.F. Pun, B. Jiang, K.S. Chin, (2003). A fuzzy multi-criteria decision support procedure for enhancing information delivery in extended enterprise networks, *Engineering Applications of Artificial Intelligence*, 16, 1–9.

[28] Shu-Hsien Liao, and Kuo-Chung Lu (2002). Evaluating anti-armor weapon using ranking fuzzy numbers, *Tamsui Oxford Journal of Mathematical Sciences*, 11(1), 33–48.

[29] M. Haghighi, A. Divandari, M. Keimasi (2010). The impact of 3D e-readiness on e-banking development in Iran: A fuzzy AHP analysis, *Expert Systems with Applications*, 37(6), 4084–4093.

[30] Anderson, W., Jr., Cox, J. E. P., & Fulcher, D. (1976). "Bank selection decisions and marketing segmentation". *Journal of Marketing*, 40(1), 40–45.

[31] Arshadi, N., Lawrence, E. C. (1987). An empirical investigation of new bank performance. *Journal of Banking and Finance*, 11(1), 33–48.

[32] Ashton, C. (1998). Balanced scorecard benefits Nat West Bank. *International Journal of Retail and Distribution Management*, 26(10), 400–407.

[33] Athanassopoulos, Giokas, D. (2000). On-going use of data envelopment analysis in banking institutions. *Evidence from the Commercial Bank of Greece. Interfaces*, 30(2), 81–95.

[34] Bauer, P. W., Berger, A. N., Ferrier, G. D., & Humphrey, D. B. (1998). Consistency conditions for regulatory analysis of financial institutions: A comparison of frontier efficiency methods. *Journal of Economic and Business*, 50(2), 85–114.

[35] Beccalli, A. (2007). Does IT investment improve bank performance? Evidence from Europe. *Journal of Banking & Finance*, 31, 2205–2230.

[36] Caballero, R., Cerda, E., Munoz, M.M., Rey, L. (2004). Stochastic approach versus multi objective approach for obtaining efficient solutions in stochastic multi objective programming problems. *European Journal of Operational Research*, 158, 633–648.

[37] Changchit, C., Terrell, M.P. (1993). A multi-objective reservoir operation model with stochastic inflows. *Computers and Industrial Engineering*, 24(2), 303–313.

[38] Chen, S.J., Hwang, C.L. (1991). *Fuzzy Multiple Attribute Decision Making*. Springer-Verlag, Berlin.

[39] Y. Y. Jusoh, K. Chamili, N. C. Pa, and J. H. Yahaya (2014). Open Source software selection using an analytical hierarchy process (AIIP), *American Journal of Software Engineering and Applications*, 3(6), 83–89.

[40] Vinay S, S. Aithal and S. Adiga (2014). Integrating goals after prioritization and evaluation a goal-oriented requirements engineering methods, *International Journal of Software Engineering & Applications (IJSEA)*, 5(6).

[41] Sumeet Kaur Sehra, Yadwinder Singh Brar and Navdeep Kaur (2016). Application of multi-criteria decision making in software engineering, *International Journal of Advanced Computer Science and Applications*, 7(7).

[42] Mary, S.A.S.A. and Suganya, G. (2016). Multi-criteria decision making using electre. *Circuits and Systems*, 7, 1008–1020. http://dx.doi.org/10.4236/cs.2016.76085.

[43] Vanessa B.S., et al. (2016). A multicriteria approach for selection of agile methodologies in software development projects, *2016 IEEE International Conference on Systems, Man and Cybernetics, SMC 2016, October 9–12, Budhapest, Hungary*.

[44] E. N. Madi, J. M. Garibaldi, C. Wagner, An exploration of issues and limitations in current methods of TOPSIS and fuzzy TOPSIS, *2016 IEEE International Conference on Fuzzy Systems (FUZZ), 978-1-5090-0626-7/16/$31.00 ©2016 IEEE*.

[45] Khan, J. (2016). Requirements prioritization using analytic network process (ANP). *International Journal of Scientific & Engineering Research*, 7(11).

[46] R. Santos, A. Albuquerque, P. R. Pinheiro (2016). Towards the applied hybrid model in requirements prioritization. *Procedia Computer Science*, 91, 909–918.

[47] E. Elsherbeiny, Hadeel, Ahmed, Abd El-Aziz, R. Nagy (2017). Decision support for requirements prioritization using data analysis. *Egyptian Computer Science Journal (ECS)*, 41.

[48] Q. Raneem, A.-S. Alaa, H. Q. Mais, H. Amjad (2017). *Requirements Prioritization Techniques Review and Analysis*, pp. 258–263. 10.1109/ICTCS.2017.55.

[49] Hassan, A. and Ramadan, N. (2017). A Fuzzy approach for Wieger's method to rank priorities in Requirement Engineering. *CiiT International Journal of Fuzzy Systems*, 9, 189–196.

[50] Hassan, A. and Ramadan, N. (2018). A proposed hybrid prioritization technique for software requirements based on fuzzy logic. *CiiT International Journal of Fuzzy Systems*, 10, 45–52.

13 Optimization Software Development Plan

Anita Soni
IES University

Prashant Richhariya
Technocrats Group of Institutions

CONTENTS

13.1   INTRODUCTION

Due to the progressive development stage of large systems, there are schedule constraints and the number of features requested typically exceeds the available resources. Software evolution [1] is the demand of today's environment; software systems must be continually adapted. It is the demand of present tendency to evolve because of the need to extend the functionality of the system by adding new features (or requirements) could represent customer wishes derived from perceived market need, or product requirements that the company developing the product consider worthwhile to pursue or correcting errors that are discovered during operation of the software. Most of the features originate from diverse stakeholders. If we opt old methods of software, then we will lack irrespective of the degree of success of an operational system, it has stakeholders that who require their needs to be met despite resource and risk constraints [23].

Incremental software development approach allows customers to receive parts of a system early – a situation that allows for creating early value; addressing this problem by allowing compromises of providing different features at different release points offers sequential releases of software systems with additive functionalities in each increment. Thus, each increment is a collection of features that form a complete system that would be of value to the customer. A major problem faced by companies to detected and to fix defects in developing or maintaining large and complex systems is deciding the features should be used releases of the software [4], considering all features of the next step of release [3].

It's critical to debunk the myth that release planning is a magic pill that guarantees "everything" will be done on time. In reality, release planning isn't about making sure that all of the work that has been scoped is performed. Rather, it's about ensuring that work is effectively prioritised, with the most important items (as determined by the PM or PO) at the very top of the backlog, and that each release achieves the required results.

Release planning is simply the practice of connecting the product's strategy – determining what desired outcomes we want to drive through one or more releases – with tactics – balancing the work to be done with constraints such as capacity, deadlines and budget while enabling progress-monitoring practices – to ensure that the product being built is evolving in the right direction. This allows you to make well-informed product decisions, optimise validated learning [5], organise how to give the most value and set realistic product expectations.

Release planning is a task that frequently brings agile teams, stakeholders and subject-matter experts together, and everyone involved should work closely together. The effectiveness and results of release planning are usually ensured by a Product Manager or a Scrum Product Owner. We've come up with a few critical criteria to bear in mind to support the PM/journey POs for effective release planning and help teams navigate their difficulties.

Release plans sit in the project management hierarchy, product roadmaps. The planning onion is a term used in agile methods to describe a framework that moves down a succession of layers, from strategic to tactical. Agile practitioners may use somewhat different terminology. The product vision, roadmap, release plan, sprint plan and daily stand-up are all part of the fundamental framework [6].

A typical scenario is when a team has a defined delivery date for a big release in a month that includes a number of items that have been meticulously recorded in the backlog. What if the PM or PO insists on including ALL of those features in the next release? Is the team capable of delivering on that promise? We don't know at this stage, but it's a high-risk strategy to try to fix both a deadline and the scope at the same time. It makes far more sense either to set a definite deadline and see how many features can be done within that time frame, or to set a fixed scope and estimate a delivery date for everything. In any event, there are a slew of interfering circumstances. Agile teams can use a collection of strategies and tools based on empirical data to make reasonably accurate predictions within their specific restrictions and conditions.

Roadmaps show a longer-term perspective, including multiple releases and sometimes even multiple projects.

Being strategic tools, they aim to capture the product vision. They communicate product and release goals and present high-level features and product capabilities [7].

Release plans are shorter-term and decidedly more granular. They're more tactical than roadmaps focusing on specific work to be done and showing details down to the level of individual backlog items (Figure 13.1).

At the iteration level, this is because there is more certainty and clarity about what features will be completed and potentially released at this point. Even though Scrum pushes for delivering potentially releasable increments by the conclusion of each sprint, it is ultimately up to the Product Owner to decide whether to release the increment.

It's also worth mentioning that Scrum prefers smaller batch sizes and more regular releases than large and infrequent releases. The reason for this is that as the number of features in a release grows, the complexity of the release grows as well, and the learning process slows dramatically because of the longer feedback cycle. Release burndown charts are a great method to keep track of your progress. Then, as time goes on, you can use that information to feed and enhance your product roadmap.

13.2   START WITH THE PRODUCT VISION AND HOW IT CAN BE REPRESENTED IN A ROADMAP

Establishing clear, defined and, most importantly, quantifiable goals is the first big step towards effective release planning. All these specific strategic objectives can be written out on a product roadmap to guide your efforts.

We favour the goal-oriented (GO) product roadmap built, despite the fact that there are other good tools for developing effective product roadmaps.

You've created a strategic product roadmap by now. You've already figured out what your goals are for one or many more releases, and you have a rough notion of which features will help you get there. But, to take it a step further, prioritizing goals and features in a rational manner will be beneficial. Identified which ones are the most important for your product's success, for example, and ensuring that your efforts are always focused on delivering the next most value thing.

Estimations can start by bringing together the relevant group of individuals – subject matter experts, architects, product specialists, business analysts and the actual product development team – to try to draw on previous experiences with similar projects and map out high-level estimates.

Estimation approaches strengthen the overall learning cycle for product teams by allowing them to compare their initial perceptions of the amount of work needed to complete a feature to the actual effort required.[8].

- Are you falling short of the proposed goal(s)?
- Why isn't your movie coming out when you want it to?
- Are you going over your budget for the release?

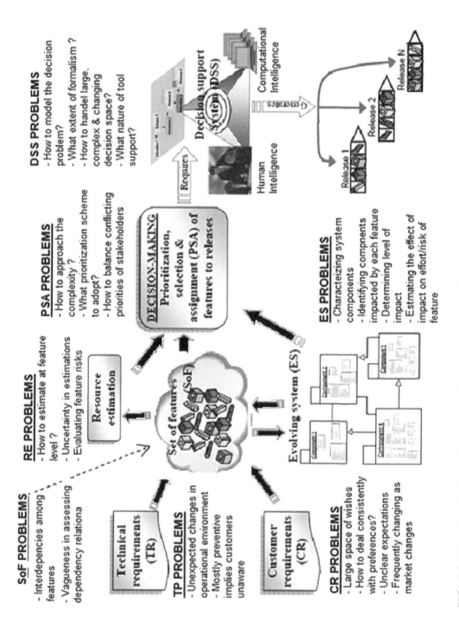

FIGURE 13.1   Instantiation of release planning as a decision problem.

Both the product roadmap and the release strategy are important to a project's success.

It's critical to keep the two perspectives aligned as time passes and the inevitable changes occur. A shift in a product's general strategy, as reflected in the product roadmap, usually always means a shift in the priorities for the features scheduled for delivery.

Simultaneously, issues at the release level, such as delays in working through the backlog, would reverberate through the planning process, eventually affecting the product roadmap.

The reason for this is simple: Even while a team should ideally be able to complete all the work stated within the requisite schedule and budget, this is not always the case. Developing a valuable product isn't a straight line, and things go wrong along the way. Market conditions or priorities may change, causing our assumptions to change.

13.3   WHAT IS INVOLVED IN RELEASE PLANNING?

Release planning will assist us in comprehending a clear product vision and a set of measurable goals defined and generally prioritised. The next important step is to get a ballpark estimate of the cost – actual budget or raw resource allocation – connected with each planned task. It might provide a fresh perspective on where priorities should be set depending on our wishes, making it easier to plan and implement.

Successful teams figure out how to reuse verified learning from prior releases to restructure the horizon of their release plan accordingly and guarantee that all aspects are [9] convergent toward achieving long-term customer pleasure.

13.4   ENHANCE COLLABORATION AND COORDINATION

13.4.1   Reduce Dev/Test Cycle Time

A faster access to test environments by the QA team reduces cycle time. Plutora notifies test teams when a new build is available from development, improving test team responsiveness and reducing time spent waiting for new code to be ready for testing (Figure 13.2).

FIGURE 13.2   Reduce/test cycle time.

13.4.2   Monitor Quality of Release Pipelines

With Plutora, release managers can quickly view test status and results in real time to continually monitor product quality and evaluate schedule risk at each phase of the CD pipeline [10] (Figure 13.3).

13.4.3   Ensure Accurate Test Coverage

As dev team velocity increases, test teams struggle to track change requests associated with new builds. Plutora automatically links change IDs with each new build, so test teams can quickly identify and assign test cases (Figure 13.4).

13.4.4   Get Insights and Reporting

Centralized dashboards provide visibility of multiple release trains across the portfolio drill-down to gain deeper insights into automated and manual test results and defect rates.

Easily Release Planning isn't a one-time process that you perform once and then forget about. For it to be effective, you need to commit to an iterative and incremental

FIGURE 13.3   Monitoring quality pipeline.

FIGURE 13.4   Accuracy testing device.

FIGURE 13.5 Reporting/output device.

approach [11]. Teams enforcing good release practices should be aware that Release Planning unfolds on two different levels (Figure 13.5).

13.5 APPLICATION OF RELEASE PLAN

- **Defining the High-Level Scope** – Ensure that all team members, at the very least, have a clear understanding of the scope.
- **Defining and Clarifying Goals** – Assist in aligning product goals with the needs of both the company and the target audience.
- **Making Rough Estimates** – The development team evaluates the expected workload during release planning.
- **Managing the Implementation Scope** – It's easy to keep track of the overall list of features and when each one must be published when we view the whole list.
- **Identifying the Constraints of the Project Triangle** – The time and budget constraints of the product development process must be factored into your strategy.

This is more strategic in nature at the Roadmap level. You're usually thinking about the outcomes of previous releases and the lessons learned from them, as well as coordinating what future releases might look like and if one or several iterations are required. Road mapping sessions that include all interested stakeholders, as well as subject-matter experts, are strongly recommended, as previously noted.

Hopefully, these principles will assist you with a foundation for determining the most successful way to organise releases for your teams. Disciplined practice that adheres to the underlying beliefs can help you achieve better results [12]. You may ensure that you're tackling it in the safest possible way by implementing only a few modifications at a time, allowing your employees to integrate them as good and long-lasting changes in their micro-culture.

13.6 WORKING PLAN RELEASES MORE EFFECTIVELY

The fundamental goal of release planning is to determine the next set of essential market features and set a release date for them.

Release planning should be a collaborative effort involving the entire development team, using members' experience and gaining buy-in for the strategy [12].

13.6.1 EXAMINE YOUR CURRENT RELEASE MANAGEMENT PROCEDURE

Begin by analysing your present release management process's personnel, processes, and tools. Capable personnel, a well-defined and regular process and a toolset that supports all participants in the process are all characteristics of a good release management role.

13.6.2 CREATE A CORPORATE RELEASE PLAN

Establishing an enterprise release plan that clearly articulates regular release cycles is also critical. It's vital to establish your release management objectives and goals early on. This can be accomplished by informal policies or a more formal way.

Goals can be defined as measures that focus on one or more of the following [12]:

* The number of releases that have been successful.
* Reduced release-related outages and downtime.
* Tracking and increasing the top line by a certain amount.
* The number of releases that were implemented late.
* The quantity of major and minor releases.
* The number of occurrences resulting from releases.
* The number of releases that have failed.
* The number of releases that were put into place, but never tested.
* By release type, the best and worst times to implement.

13.6.3 DEFINE THE OPTIMAL RELEASE MANAGEMENT PROCESS

To begin, identify the inputs to the release management process, such as portfolio and program management systems, service management systems, quality management systems, configuration management systems and deployment solutions.

Second, identify essential tasks, including release planning, coordination, design, build, and configuration of releases, release acceptance coordination, rollout planning, coordination of deployment to production and performance assessment against key criteria [13].

Third, determine the outputs of release management, such as incident management, change management, service level management and service monitoring.

13.6.4 PUT MONEY INTO THE APPROPRIATE INDIVIDUALS

In a good release management process, the release manager, environment managers, test managers and implementation managers all play important roles. Program and project managers, on the other hand, oversee a wide range of workflows and operations to meet crucial deadlines. Developers are managed by development managers, who create work packages for deployment [14].

Leadership, organisation and planning, as well as technical depth, project management, communication and teamwork, are all necessary in these roles.

13.6.5 Make Use of the Appropriate Tools

You're presumably utilising a mix of tools for development, testing and operations. You'll also need a powerful release management solution that can aid with stakeholder management, communication, a master release schedule, automated workflow features, dashboards with reports and the ability to interact with your existing toolset.

13.6.6 Make the Most of the Testing Environment

IT environments must be set up for test execution and validation at all phases of the release process. Hardware, storage, network connections, bandwidth, software licencing, user profiles and access rights are all part of the release infrastructure. To minimise any environment bottlenecks, it is vital to understand dependencies and reduce contention.

13.6.7 Define Stages and Activities to Govern

At a physical level, work packages are promoted via numerous environments for various forms of testing and validation as releases progress through their major phases, integrated gates and milestones. As a result, significant rework is avoided by having a transparent baseline of the environments and a clear understanding of the composition of work packages.

13.6.8 Ensure Stakeholder Engagement is Transparent

Set release dates and encourage your team to strive towards not only the ultimate release, but also interim goals such as integrated testing completion. Engage stakeholders to prioritise unresolved feature requests and allocate them to future releases once the release dates have been determined and agreed. Customers gain delivery trust from regular, controlled releases.

13.6.9 Make Ongoing Communication Possible

As far as feasible, make sure that information on the release's progress is available in a frictionless manner. To put it another way, all parties should have a system of record that allows them to obtain the data they require in real time.

13.6.10 Keep an Eye on the Numbers

Monitor end-to-end release health by tracking key indicators on a regular basis. To drive your team to meet and exceed objectives, it's critical that they understand the business value of your release management role.

13.7 CONCLUSION

You must increase testing predictability and prediction accuracy if you are to satisfy your business commitments. Implement watertight releases and achieve your goals, starting with a status quo evaluation and ending with metric measurement.

The ideal approach, however, is to use a planning tool such as Plutora, which allows you to document and validate deployment plans, staging and rollout – reducing risk and increasing ROI.

REFERENCES

1. Saliu, M., and Ruhe, G., (2005). Software release planning for evolving systems. *Innovations in Systems and Software Engineering*, 1, 189-204, 2005. Doi: 10.1007/s11334-005-0012-2.
2. Natt-och-Dag, J., Regnell, B., Gervasi, V., and Brinkkemper, S., A linguistic- engineering approach to large-scale requirements management, *IEEE Software*, 22, 1, 32–39, 2005.
3. Ahmed, M., Saliu, M. O., and AlGhamdi, J., Adaptive fuzzy logic-based framework for software development effort prediction, *Information and Software Technology*, 47, 1, 31–48, 2005.
4. Bagnall, A. J., Rayward-Smith, V. J., and Whittley, I. M., The next release problem. *Information and Software Technology*, 43, 14, 883–890, 2001.
5. Carlshamre, P., Release planning in market-driven software product development: Provoking an understanding, *Requirements Engineering*, 7, 3, 139–151, 2002.
6. Ruhe, G., Software release planning, In *Handbook of Software Engineering and Knowledge Engineering*, vol. 3, Chang, S. K. (Ed.): World Scientific Publishing, 2005.
7. http://www.lucidchart.com/documents/editNewOrRegister/9da7fb63-6026-4521-b8b3-3a81645e71d2.
8. Fisseha, B., *Data Scientist*. https://www.linkedin.com/in/fisseha-berhane-phd-543b5717?trk=nav_responsive_tab_profile.
9. ScienceDirect. The RO method is defined as an alternative modelling method for handling optimization problems with uncertain parameters. In *Classical and Recent Aspects of Power System Optimization*, 2018.
10. Blake, S., *How to Approach Your Agile Release Plan for Successful Development*, Head of Marketing, Wollongong.
11. https://www.knowledgehut.com/tutorials.
12. Lehman, M. M., Laws of software evolution revisited. In *Proceedings of 5th European Workshop on Software Process Technology (EWSPT'96), Nancy, France*, 1996, pp. 108–124.
13. Denne M, Cleland-Huang J., The incremental funding method: data driven software development. *IEEE Software*, 21, 3, 39–47, 2004.
14. Saliu O, Ruhe G., Supporting software release planning decisions for evolving systems. In *Proceedings of 29th IEEE/NASA Software Engineering Workshop*, Greenbelt, USA, 6–7 April, 2005.

14 A Time-Variant Software Stability Model for Error Detection

Saurabh Sharma
Amity University

Ashish Mishra
Gyan Ganga Institute of Technology and Sciences

Harish K. Shakya
Amity University

CONTENTS

14.1 INTRODUCTION

Several transportation infrastructure failures have been blamed for extreme climate-related events (such as floods and storms) in recent decades. Several studies, including Muis et al. (2015), Winsemius et al. (2016) and Wang et al. (2018a), predict a significant rise in future flood dangers, which they ascribe to climate change. Changes in temperature profiles, precipitation patterns, sea level and the frequency of coastal storms are only a few of the consequences (Neumann et al. 2015).

Climate change, according to Arnell and Gosling (2016), might result in a more than 180% increase in global flood risk by 2050. As a result, unless new infrastructure management methodologies capable of accounting for this change are adopted, transportation structures' susceptibility and failure risk may grow significantly.

Bridges should be a key focus of these management strategies due to the potentially devastating and debilitating repercussions of their failure.

DOI: 10.1201/9780367816414-14

Accelerated scour, erosion of bridge approaches and high loads due to direct water pressure and debris impact are just a few of the variables that might cause a bridge to collapse partially or completely during floods (Ettouney and Alampalli 2011). By lowering the buckling resistance and lateral capacity of pile foundations, scour can jeopardise the stability of a shallow foundation. Bridges will be more vulnerable to future floods or other catastrophic events such as seismic excitations or traffic overloads as a result of these effects (Hung and Yau 2014; Banerjee and Ganesh Prasad 2013; Ganesh Prasad and Banerjee 2013). To effectively assess bridge dependability during severe occurrences, a comprehensive technique capable of assessing bridge performance under projected hazard intensities should be applied. In recent decades, fragility models have gained significant acceptance among infrastructure managers as a useful tool for analysing the operation of facilities exposed to natural disasters (e.g. earthquakes and hurricanes). Given a set degree of danger, a fragility model calculates the risk that a structure will achieve or surpass a defined damage condition (Gidaris et al. 2017). Many types of vital infrastructure, such as nuclear power plants and dams, are subjected to these models in order to assess their seismic risk. They're also utilised to assess bridge performance in the face of earthquakes (Wang et al. 2014a), tsunamis (Akiyama et al. 2012), hurricane-induced surge and wave hazard (Ataei and Padgett 2012) or the combined impact of many hazards (Ataei and Padgett 2012; Wang et al. 2014b; Banerjee and Ganesh Prasad 2013). Despite the fact that river flooding is responsible for 28% of bridge collapses in the USA (Cook et al. 2013), river flood bridge fragility models are sparse (Gidaris et al. 2017).

The Hazus (2018) approach uses data from the National Bridge Inventory database (FHWA 2016) to calculate empirical failure probability as a function of flood return duration and scour vulnerability rating, making it one of the few flood fragility models available in the literature. Failure is defined as the presence of damage that costs 25% of the bridge's replacement cost since there aren't enough data to calibrate the model (Hazus 2018). However, such qualitative models may not be accurate enough to be employed in infrastructure management because no substantial structural analysis is generally conducted. Turner (2016) used hydrodynamic uplift forces as the primary failure criterion to develop fragility curves for a number of Colorado bridges. The results were used to determine how much elevation adjustment was required to increase the bridge's resistance to hydrodynamic uplift forces.

Other flood-related failures, such as pier failure due to scour or horizontal water pressure, were left out. A probabilistic examination of gauge station records in the research region was also used to calculate flood frequency. However, true danger occurrence probability may fluctuate dramatically as a result of climate change (Arnell and Gosling 2016; Khandel and Soliman 2019). When assessing failure probability, Kim et al. (2017) developed a flood fragility model for bridges that takes into consideration bridge scour, structural degradation and debris build-up. In their study, they employed finite element (FE) analysis and reliability estimates. Due to the computational costs associated with probabilistic analysis involving FE modelling and Monte Carlo simulation, a simplified FE model was incorporated into their computational approach, and the first-order reliability method (FORM) was used to compute the failure probability under a limited number of random parameters.

To facilitate the application of FE analysis in probabilistic simulations while keeping an acceptable processing cost, some researchers employ approximation approaches such as response surface analysis to build an analytical link between the structural response and the underlying variables (e.g. Buratti et al. 2010; Park and Towashiraporn 2014). The link may then be tested using Monte Carlo simulation or other standard dependability methodologies such as FORM. Response surface and FORM techniques, on the other hand, might suffer from a lack of accuracy when dealing with highly nonlinear problems or when several failure modes must be addressed (Kroetz et al. 2017; Song et al. 2018; Wang et al. 2018b). As a result, sophisticated surrogate modelling approaches, including polynomial chaos expansion (PCE), kriging models and artificial neural networks (ANNs), can help in mimicking the behaviour of complex and nonlinear structural systems with many failure modes. Approaches based on ANNs have been shown to converge quicker and yield a shorter computation time for difficult functions when compared to PCE and kriging models (Kroetz et al. 2017).

ANNs are sometimes referred to as "black-box" systems, suggesting that the majority of their parameters are unknown (Zhang et al. 2002). Surrogate modelling approaches, on the other hand, such as local Gaussian processes, polynomial response surfaces, support vector machines and kriging models, approximate the response function without requiring a physical understanding of the system processes (Ferrario et al. 2017). Given the complexity of the functions that ANNs are supposed to mimic, completely training these models may need a large number of data. ANNs, on the other hand, can be used in conjunction with adaptive experimental design techniques to minimise the number of training samples required (de Santana Gomes 2019). The use of contemporary and effective optimisation algorithms, as well as the availability of a large number of cloud computing resources for machine learning applications, can aid in the management of the computational costs associated with these models.

Computationally efficient approaches are necessary to appropriately integrate extensive FE modelling in the fragility study of bridges under flood hazard. This technique should also analyse the entire collection of random variables linked to bridge resistance, load effects and hazard occurrence likelihood in light of changes expected to occur as a result of long-term variability in climatic trends. This study fills that need by proposing a probabilistic technique based on deep learning neural networks for analysing the time-variant fragility of bridges under floods and flood-induced scour while taking into consideration future climatic unpredictability. The proposed technique uses downscaled global climate modelling data to predict future time-dependent scour patterns under various climatic scenarios. A deep learning (DL) algorithm (DN 1) is utilised throughout the basin to estimate streamflow using anticipated precipitation and temperature profiles. The expected streamflow profiles are then used in a probabilistic simulation to calculate the long-term scour depth and flood threat. A FE model is used to create the data set needed to train a second DL network (DN 2) capable of predicting the behaviour of the bridge foundation during flood and flood-induced scour. The effects of long-term material deterioration (i.e. corrosion) are taken into account. After that, a Monte Carlo simulation is utilised to evaluate failure probability and generate a fragility surface using the second trained

DL network (DN 2). River discharge is the hazard intensity metric, and the fragility surface depicts the risk of bridge collapse over a certain service life.

14.1.1 Analysing Climate Data

Over the last few decades, a lot of scientific work has gone into investigating climate behaviour and anticipating future climate patterns (e.g. Sheffield et al. 2013). Now in its fifth phase, the Coupled Model Intercomparison Project (CMIP5) is a cutting-edge tool for acquiring a complete picture of past and future climate patterns (Taylor et al. 2012). More than 50 different models capable of assessing past and future climate are included in the CMIP5 data collection. The models differ in terms of model formulations, experiment conditions, climate noise and model resolutions. Multi-model ensembles are also employed to mitigate the effects of model uncertainty (Taylor et al. 2012). Due to the considerable processing costs associated with dependability analysis under climate change, using all available climate models may not be feasible. Furthermore, because not all climate models can generate correct results in every location, global climate models (GCMs) should be carefully selected.

Climate data from GCMs may be compared to historical records to determine whether GCMs are appropriate for a certain location (Samadi et al. 2010).

Future greenhouse gas (GHG) emission scenarios are another key source of uncertainty in climate prediction. Radiative forcing patterns characterise emission scenarios in modern climate modelling practice (Moss et al. 2010). Representative concentrative pathways characterise radiative forcing, which is defined as the difference between absorbed insolation energy and radiation energy reflected by the Earth (RCPs). RCP 2.6, RCP 4.5, RCP 6.0 and RCP 8.5 are the four most common RCP instances. Several RCPs may be used to account for variations in GHG emissions and concentration paths, as well as land use and future land cover (Shrestha et al. 2016).

GCM outputs are frequently constructed at high spatial resolutions (125–500 km grids). Because hydrological impact studies require fine-resolution data (typically 10–30 km grids), the GCMs' coarse resolution will not be suitable for regional-scale estimates (Frost et al. 2011). Coarse-resolution data can be transformed to fine-resolution data using dynamic or statistical downscaling methods. The daily bias correction constructed analogues (BCCA) downscaling approach is used in this work (Maurer et al. 2010). This approach is a hybrid statistical strategy that uses both quantile mapping bias correction and daily downscaling processes to conduct downscaling. The hybrid performance of this model, according to Maurer et al. (2010), results in exceptionally accurate climate forecasts at regional scales. Multiple downscaling procedures should be investigated, with the most relevant ones for the area of interest being integrated into climate projections to account for downscaling uncertainty (McPherson 2016). With the addition of different climate modelling variables, each with its own GCM, downscaling procedures and RCP values, multiple climatic scenarios may be created. In this analysis, 18 climatic data sets were used, including 3 different GCMs, 3 RCP scenarios and 2 ensemble runs of each model. For each combination of climate model and emission scenario, there are several ensemble runs with different initial condition assumptions. The model's predictions differ somewhat based on the starting conditions. The influence of the models' internal variability (i.e.

under various beginning circumstances) fades considerably in the long run. This is especially true when contrasted to other forms of uncertainty, such as model uncertainty (multiple models) and scenario uncertainty (different future emissions pathways). Several studies (e.g. Hawkins and Sutton 2009; Yip et al. 2011) have delved deeper into assessing the impact of various uncertainties on climate forecasting.

14.1.2 PREDICTION OF LONG-TERM PIER SCOUR

Flood-induced scour can have a significant influence on flood-prone bridges' time-variant strength and stability. In the literature, there are numerous scour depth prediction techniques (e.g. Breusers et al. 1977; Briaud et al. 2001). These formulas, which are mostly based on flume test tests, can account for the impacts of pier size, shape and alignment on the maximum estimated scour depth. Erosion occurs on a particle-by-particle basis in cohesionless soils (Arneson et al. 2012).

The pace of scour initiation is increased to the point that the maximum scour depth is attained in a couple of hours or a few flood events. Cohesive soils, on the other hand, rely heavily on electromagnetic and electrostatic interparticle interactions, resulting in a slower rate of scour (Arneson et al. 2012).

To evaluate the rate of scour in various soil types, an erosion function apparatus (EFA) test can be utilised (Briaud et al. 2001). This test determines the equivalent time (t) necessary to erode 1 mm of soil at various flow velocities (v). The erosion rate [Z 14 1 $=t$ in mm $=h$] and the hydraulic shear stress acting on the soil () are computed based on the results of the EFA test.

The maximum pier scour depth (Z_{max}) is calculated by (Arneson et al. 2012)

$$Z_{max} = 2.0\lambda_1 y_1 K_1 K_2 K_3 \left(\frac{a}{y_1}\right)^{0.65} Fr_1^{0.43} \qquad (14.1)$$

where 1 is the modelling uncertainty factor, y_1 is the flow depth upstream of the pier, K_1 is the pier nose shape correction factor, K_2 is the angle of attack correction factor, K_3 is the bed condition correction factor, $a=$ pier width, and Fr_1 is the Froude number given by

$$Fr_1 = \frac{V}{\sqrt{(gy_1)}}$$

where V is the mean velocity of the river directly upstream of the pier and g is gravity's acceleration (9.81 m $=s_2$). The scour depth (Z) is calculated as a function of time (Briaud et al. 2001).

14.1.3 PREDICTION OF STREAMFLOW AND FLOODS

In this work, the results of global climate modelling are utilised to estimate future flood threats. GCMs give climate-related metrics such as anticipated precipitation and temperature profiles, but flood prediction using these elements is problematic.

Estimating river discharge using precipitation and temperature profiles involves detailed hydrologic modelling of the basin. Such a thorough hydrological examination may need a substantial amount of resources or instruments, which infrastructure managers may lack. This process may be sped up by employing cutting-edge computing techniques such as machine learning, resulting in computationally efficient, but highly accurate streamflow forecasts. In this work, downscaled temperature and precipitation data from multiple climate scenarios are utilised to anticipate future river flow patterns using TensorFlow.

It's worth noting that many statistical or hydrological streamflow forecasting methods assume stationary circumstances (Humphrey et al. 2016).

Changes in channel flow morphology and precipitation patterns compared to previous data can all produce nonstationarity (Westra et al. 2014). The streamflow modelling method used in this article, which employs DL neural networks, assumes stationary parameters. The DL network that was utilised to estimate future river discharge, designated as DN 1, will be addressed in greater depth later in this chapter.

14.1.4 FLOOD AND FLOOD-INDUCED SCOUR BEHAVIOUR OF BRIDGE FOUNDATIONS

This study looks on the stability of bridges with deep foundations. Several strength and serviceability limit states are examined to evaluate the time-variant dependability of a foundation under horizontal and vertical loads, and OpenSees FE software (Mazzoni et al. 2006) is used to model the piling group's reaction under applied stresses. Nonlinear springs and displacement-based beam-column components are used to represent the piles. In addition to the aforementioned elements, pile nodes, fixed spring nodes and slave spring nodes are given. The pile elements are replicated by the beam-column elements, while the springs, which are made up of zero-length elements in both horizontal and vertical orientations, approximate soil behaviour.

P-y springs (API 1987) are used to mimic lateral soil behaviour, whereas t-z springs (Mosher 1984) and q-z springs (Vijayvergiya 1977) are used to model shaft and tip behaviour, respectively. The internal friction angle (), unit weight () and soil shear modulus () are all used to define the springs (G). Every node is three-dimensional, with six degrees of freedom for rotation and transition. The pile components, fixed springs and slave nodes are distributed vertically along the length of the embedded pile. The embedded length of the piles is adjusted in response to the scour depth forecasts (i.e. pile embedded length $L2 = $ total pile length − scour depth Z). A schematic of the FE model that was employed is shown in Figure 14.1.

When analysing the behaviour of closely spaced pile groups under lateral and axial pressures, the effects of pile–soil–pile interactions must be taken into account. These impacts typically result in a decline in soil resistance, which may be alleviated by altering the reaction of particular heaps in the right way (Brown and Reese 1988). Updated p-y curves accounting for group effects may be generated for laterally loaded piles by applying the reduction factors to the p-values (Dunnavant and O'Neill 1986). Furthermore, efficiency factors can be exploited to alter pile group behaviour under axial stresses (O'Neill 1983). Based on Dunnavant and O'Neill's

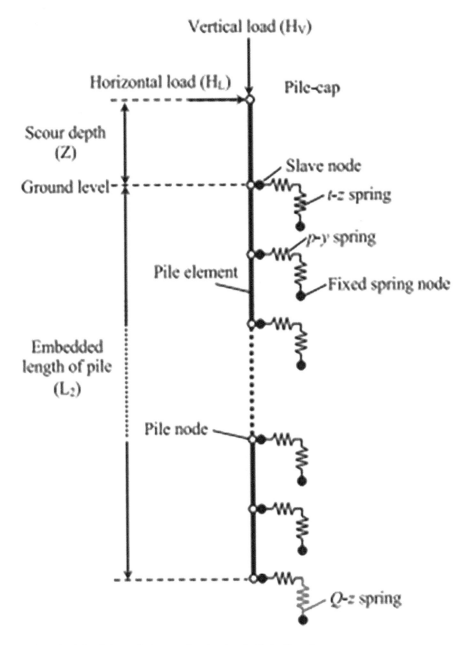

FIGURE 14.1   Schematic layout of a simulated pile in OpenSees.

experimental work, resistance reduction factors accounting for group effects in the lateral direction are developed in this study (1986). The training data set for a second TensorFlow DL neural network, DN 2, is provided by the constructed FE model. A complete factorial experimental design (Dieter 2000) is used to create an inclusive training data set that encompasses the conceivable range of input parameters encountered in the subsequent analytic processes. After that, a probabilistic analysis is conducted to assess bridge fragility under flood loads using the trained DL network (DN 2).

14.1.5 Effects of Long-Term Corrosion

Under lateral and axial stress conditions, steel pile corrosion can result in a loss in capacity. A typical source of this decline is the loss of section thickness due to severe environmental conditions or repeated dry-wet cycles (ElGawady et al. 2019). Corrosion losses are influenced by moisture, sulphate, chloride and microbial concentration. The resistivity, chemical composition and pH of the soil, as well as the location of the water table and oxidation potential, all impact corrosion propagation in steel piles (Ding 2019). The rate of corrosion in soil decreases over time, according to the literature (e.g. Schlosser and Bastick 1991). This is due to the depletion of oxygen and the formation of a protective layer from corrosion products (Ohsaki 1982).

As a result, the corrosion rate may be expressed as the following time-dependent relationship (Kucera and Mattsson 1987).

14.1.6 Conclusions

A probabilistic framework for assessing bridge fragility under flood and flood-induced scour in the context of climate change is presented in this study. Climate data sets for the location, downscaled precipitation and temperature were extracted from the CMIP5 archive.

These variables are used to forecast river discharge and scour depth.

A deep feedforward neural network (DN 1) was utilised to predict discharge, and it was trained using historical data at the bridge site.

An OpenSees FE model generated the necessary training data for a model.

A second deep neural network (DN 2) is employed to compute the internal state.

Given the foundations' service life, pressures and displacements and river discharge, the yearly failure probability of the bridge was calculated and utilised in flood-prone areas. The time-varying fragility of the bridge's surfaces. The proposed approach may be utilised to produce a bridge flood, according to the findings.

Fragility rises as a function of service life and river discharge. Officials at the bridge can make educated judgments based on the fragility of the surface. Decisions on management activities (e.g. retrofit) are taken with the goal of reducing the chance of failure in the case of future floods. It also provides a quantitative metric that might help in the decision-making process when it comes to bridge closures during floods.

Given the temperature and precipitation patterns, the utilised neural network (DN 1) is capable of forecasting streamflow with adequate precision. This was verified by a daily coefficient of determination (R2) of 0.9116 during the calibration phase.

The results of the streamflow prediction were also presented. Based on a detailed hydrological investigation, there is a high degree of consistency with those reported in the literature.

REFERENCES

AASHTO. 2017. *Specifications for Bridge Construction (LRFD)*. AASHTO, Washington, DC, 8th ed.

Abadi, M. et al. 2016. TensorFlow: A large-scale machine learning system. In *Proceedings of the 12th USENIX Symposium on Operating Systems Design and Implementation (OSDI'16)*. The USENIX Association is based in Berkeley, California, pp. 265–283.

Ahammed, M. and R. E. Melchers, 1997. Underground pipes subjected to coupled loads and corrosion: A probabilistic analysis. *Eng. Struct.* 19(12), 988–994. https://doi.org/10.1016/S0141-0296(97)\s00043-6.

Akiyama, M., D. M. Frangopol, M. Arai, and S. Koshimura, 2012. A probabilistic analysis of bridge structural performance under tsunami threat. In *Proceedings of the 2012 Structures Congress*. ASCE, Reston, VA, pp. 1919–1928.

Ang, A. and W. Tang, 2007. *The Application of Probability Ideas in Engineering Planning and Design, with a Focus on Civil and Environmental Engineering*. Wiley, New York.

API (Application Programming Interface) (American Petroleum Institute). 1987. *Planning, Designing, and Constructing Fixed Offshore Platforms Is Recommended*. API, Washington, DC, 17th ed.

Arnell, N. W. and S. N. Gosling, 2016. Climate change's effects on river flood risk on a world-wide scale. *Clim. Change*, 134, 3, 387–401. https://doi.org/10.1007/s10584-014-1084-5.

Arneson, L. A., L. W. Zevenbergen, P. F. Lagasse, and P. E. Clopper, 2012. *Bridge Scour Is Being Assessed. Circular No. 18 on Hydraulic Engineering (HEC-18)*. Federal Highway Administration, Washington, DC.

Ataei, N. and J. E. Padgett, 2012. Probabilistic modelling of bridge deck unseating during hurricane events. *J. Bridge Eng.*, 18, 4, 275–286. https://doi.org/10.1061/(ASCE)BE.1943-5592.0000371.

Balomenos, G. P. and J. E. Padgett, 2018. Fragility analysis of pile-supported wharves and piers exposed to storm surge and waves. *J. Waterway Port. Coast. Ocean Eng.* 144, 2. https://doi.org/10.1061/(ASCE)WW.1943-5460.0000436.

Banerjee, S. and G. Ganesh Prasad, 2013. Seismic risk assessment of reinforced concrete bridges in flood-prone areas. *Struct. Infrastruct. Eng.* 9, 9, 952–968. https://doi.org/10.1080/15732479.2011.649292.

Bickel, S., M. Brückner, and T. Scheffer, 2009. Discriminative learning under covariate shift. *J. Mach. Learn. Res.* 10, Sep, 2137–2155.

Brekke, L., B. L. Thrasher, E. P. Maurer, and T. Pruitt, 2013. *Climate and Hydrological Projections from CMIP3 and CMIP5: Release of Downscaled CMIP5 Climate Projections, Comparison with Previous Data, and Overview of User Needs*. Technical Services Center, US Department of the Interior, Bureau of Reclamation, Denver.

Breusers, H. N. C., G. Nicollet, and H. W. Shen, 1977. Around cylindrical piers, local scour. *J. Hydraul. Res.* 15, 3, 211–252.

Briaud, J. L., H. C. Chen, K. W. Kwak, S. W. Han, and F. C. K. Ting, 2001. Prediction of scour rate at bridge piers using a multiflood and multilayer approach. *J. Geotech. Geoenviron. Eng.* 127, 2, 114–125. https://doi.org/10.1061/(ASCE)1090-0241.

Broding, W. C., F. W. Diederich, and P. S. Parker, 1964. Based on a reliability design criterion, structural optimization and design. *J. Spacecraft Rock.* 1, 1, 56–61. https://doi.org/10.2514/3.27592.

Brown, D. A., and L. C. Reese, 1988. *A Large-Scale Pile Group Subjected to Cyclic Lateral Loading and Its Behaviour.* Geotechnical Engineering Center, Texas University at Austin, Austin, TX.

Buratti, N., B. Ferracuti, and M. Savoia, 2010. Response surface with random factors for seismic fragility of reinforced concrete frames. *Struct. Saf.* 32, 1, 42–51. https://doi.org/10.1016/j.strusafe.2009.06.003.

Chan, C. L., and Low, B. K., 2012. Probabilistic study of laterally loaded piles using response surface and neural network techniques, *Comput. Geotech.* 43, Jun, 101–110. https://doi.org/10.1016/j.compgeo.2012.03.001.

Cook, W., P. J. Barr, and M. W. Halling, 2013. Rate of bridge failure. *J. Perform. Constr. Facil.* 29, 3, 04014080. https://doi.org/10.1061/(ASCE)CF.1943-5509.0000571.

Croke, B. F. W., F. Andrews, J. Spate, and S. M. Cuddy, 2005. *Second Edition of the IHACRES User Guide. 2005/19 Technical Report.* iCAM, School of Resources, Environment and Society, Australian National University Press, Canberra, Australia.

de Santana Gomes, W. J. 2019. Structural reliability analysis using adaptive artificial neural networks. *ASME J. Risk Uncertain. B*, 5, 4, 1–8. https://doi.org/10.1115/1.4044040.

Decker, J. B., K. M. Rollins, and J. C. Ellsworth, 2008. Analysis and prediction of pile corrosion rates based on long-term field performance. *J. Geotech. Geoenviron. Eng.* 134, 3, 341–351. https://doi.org/10.1061/(ASCE)1090-0241.

Dieter, G. 1958. *Engineering Design: A Materials and Processing Approach (McGraw-Hill Series in Mechanical Engineering).* 3rd ed. McGraw-Hill, Boston.

Ding, L. 2019. *Corrosion Behavior of H-Pile Steel in Different Soils.* Doctoral dissertation, Clemson University, Clemson. https://tigerprints.clemson.edu/alldissertations/2334.

Dunnavant, T. W. and M. W. O'Neill. 1986. Analysis of vertical pile groups subjected to lateral load using design-oriented approaches. In *Proceedings of the International Conference on Numerical Methods in Offshore Piling, Nantes, France.* Universiti Teknologi Petronas, Paris, pp. 303–316.

ElGawady, M. A., M. M. Abdulazeez, A. Ramadan, B. Sherstha, A. Gheni, E. Gomaa, Y. Darwish, 2019. *Behavior and Restoration of Corroded Steel H-Piles Phase I (Axial Behaviour).* 25-1121-0005-133-1 is the number of the report. Mid-America Transportation Center, Lincoln, Nebraska.

Ettouney, M. M. and S. Alampalli, 2011. *Civil Engineering Applications and Control of Infrastructure Health.* Taylor and Francis, CRC Press, Boca Raton, FL.

FEMA, 2011. *The Principles and Techniques of Planning, Locating, Designing, Constructing, and Maintaining Residential Buildings in Coastal Locations Are Outlined in this Document.* 4th ed. FEMA, Washington, DC.

Ferrario, E., N. Pedroni, E. Zio, and F. Lopez-Caballero, 2017. Seismic analysis of structural systems using bootstrapped artificial neural networks. *Struct. Saf.* 67, July, 70–84. https://doi.org/10.1016/j.strusafe.2017.03.003.

FHWA (Federal Highway Administration). 2016. *Dataset for the National Bridge Inventory (NBI).* On November 21, 2018, I Was Able to Access This Information. https://www.fhwa.dot.gov/bridge/nbi/ascii2016.cfm.

Frost, A. J. et al., 2011. Under Australian conditions, a comparison of multi-site daily rainfall downscaling algorithms. *J. Hydrol.* 408, 1–2, 1–18. https://doi.org/10.1016/j.jhydrol.2011.06.021.

Galambos, T. V., *International Journal of Steel Structures*, 4, 4, 223–230.

Gidaris, I., J. E. Padgett, A. R. Barbosa, S. Chen, D. Cox, B. Webb, and A. Cerato, 2017. State-of-the-art review of multiple-hazard fragility and restoration models of highway bridges for regional risk and resilience assessment in the United States. *J. Struct. Eng.* 143, 3, 04016188. https://doi.org/10.1061/(ASCE)ST.1943-541X.0001672.

Haldar, S. and D. Basu, 2014. Resistance factors for laterally loaded piles in clay, geo-characterization and modeling for sustainability. In *Geo-Congress 2014*. ASCE, Reston, VA, pp. 3333–3342.

Hannigan, P. J. F. Rausche, G. E. Likins, B. R. Robinson, and M. L. Becker, 2016. *Volume II Design and Construction of Driven Pile Foundations*, Geotechnical Engineering Circular No. 12. FHWA-NHI-16-010 is the reference number. Federal Highway Administration, Washington, DC.

Hawkins, E. and R. Sutton, 2009. The potential to reduce uncertainty in regional climate predictions. *Bull. Am. Meteorol. Soc.* 90, 8, 1095–1108. https://doi.org/10.1175/2009BAMS2607.1.

Hazus, 2018. *User Instructions for the Hazus Flood Model*. FEMA, Washington, DC.

Hosmer, D. W., Jr., S. Lemeshow, and R. X. Sturdivant, 2013. *Applied Logistic Regression*, volume 398, Wiley, New York.

Humphrey, G. B., M. S. Gibbs, G. C. Dandy, and H. R. Maier, 2016. A hybrid approach to monthly streamflow forecasting: Integrating hydrological model outputs into a Bayesian artificial neural network. *J. Hydrol.* 540, Sep, 623–640. https://doi.org/10.1016/j.jhydrol.2016.06.026.

Hung, C.-C., and W.-G. Yau. 2014. The response of scoured bridge piers to flood-induced stresses. *Eng. Struct.* 80, 241–250. https://doi.org/10.1016/j.engstruct.2014.09.009.

Jiang, S. H., S. B. Wu, C. B. Zhou, and L. M. Zhang, Li, D. Q., 2013. For structural reliability analysis with complex performance function, modelling multivariate distributions using Monte Carlo simulation. *J. Risk Reliability Proc. Inst. Mech. Eng.* 227, 2, 109–118. https://doi.org/10.1177/1748006X13476821.

Johnson, P. A., P. E. Clopper, L. W. Zevenbergen, and P. F. Lagasse. 2015. Estimating scour in bridges: Quantifying uncertainty and reliability. *J. Hydraul. Eng.* 141, 7: 04015013. https://doi.org/10.1061/(ASCE)HY.1943-7900.0001017.

Kadar, I. and L. Nagy. 2017. An investigation of the coefficients of variation and types of distributions for several soil characteristics. In *Proceedings of the 6th International Conference of Young Geotechnical Engineers (iYGEC6). The International Society for Soil Mechanics and Geotechnical Engineering publishes a Journal Called Soil Mechanics and Geotechnical Engineering.*

Karpathy, A. and L. Fei-Fei, 2015. Deep visual-semantic alignments for image description generation. In *IEEE Conference on Computer Vision and Pattern Recognition*. IEEE, New York, pp. 3128–3137.

Khandel, O. and M. Soliman, 2019. An integrated methodology for estimating the impact of climate change on the risk of bridge failure due to flooding and scour caused by flooding. *J. Bridge Eng.* 24, 9, 04019090. https://doi.org/10.1061/(ASCE)BE.1943-5592.0001473.

Kim, H., S. H. Sim, J. Lee, Y. J. Lee, and J. M. Kim. 2017. Analysis of flood fragility in bridges with numerous collapse modes. *Adv. Mech. Eng.* 9, 3, 168781401769641. https://doi.org/10.1177/1687814017696415.

Kingma, D. P. and J. Ba, 2014. *Adam: A Stochastic Optimization Approach*. The following is a preprint that was submitted on December 22, 2014. http://arxiv.org/abs/1412.6980.

Kroetz, H. M., R. K. Tessari, and A. T. Beck, 2017. Performance of global metamodeling techniques in the solution of structural reliability problems. *Adv. Eng. Software* 114, 394–404. https://doi.org/10.1016/j.advengsoft.2017.08.001.

Kucera, V. and E. Mattsson, 1987. Atmospheric corrosion. *F. B. Mansfeld's Corrosion Mechanics*, Chap. 5. Marcel Dekker, New York, 211–284.

Liaw, A. and M. Wiener, 2002. Classification and regression by random forest, *R News*, 2, 3, 18–22.

Mann, N. R., N. D. Singpurwalla, and R. E. Schafer, 1974. *Methods for Analysing Reliability and Life Data Statistically*. Wiley, New York.

MathWorks. 2016. *SIMULINK for Technical Computing*. http://www.mathworks.com.

Maurer, E. P., H. G. Hidalgo, T. Das, M. D. Dettinger, and D. R. Cayan, 2010. The value of daily large-scale climate data in assessing climate change consequences on daily streamflow in California. *Hydrol. Earth Syst. Sci.* 14, 6, 1125–1138. https://doi.org/10.5194/hess-14-1125-2010.

Mazzoni, S., F. McKenna, M. H. Scott, and G. L. Fenves, 2006. *Manual for the OpenSees Command Language. Pacific Earthquake Engineering Research (PEER) Center, 264*. Univ. of California, Berkeley, CA.

McPherson, R. 2016. *Impacts of Climate Change on Flows in the Red River Basin*. South Central Climate Science Center, Norman, OK.

Melchers, R. E., and A. T. Beck. 2018. *Structural Reliability Analysis and Prediction*. 3rd ed. Wiley, New York.

Mosher, R. L. 1984. *Load Transfer Criteria for Numerical Analysis of Axial Loaded Piles in Sand*. US Army Engineering and Waterways Experimental Station, Automatic Data Processing Center, Vicksburg, MI.

Moss, R. H., et al. 2010. The next generation of scenarios for climate change research and assessment. *Nature* 463, 7282, 747. https://doi.org/10.1038/nature08823.

Muis, S., B. Güneralp, B. Jongman, J. C. Aerts, and P. J. Ward, 2015. Flood risk and adaptation strategies under climate change and urban expansion: A probabilistic analysis using global data. *Sci. Total Environ.* 538, 445–457. https://doi.org/10.1016/j.scitotenv.2015.08.068.

Neumann, J. E., et al. 2015. Climate change risks to US infrastructure: Impacts on roads, bridges, coastal development, and urban drainage. *Clim. Change* 131, 1, 97–109. https://doi.org/10.1007/s10584-013-1037-4.

NOAA (National Oceanic and Atmospheric Administration), 2018. *Climate Data Online: Dataset Discovery*. Accessed November 10, 2018. http://www.noaa.gov/.

O'Neill, M. W. 1983. Group action in offshore piles. In *Geotechnical Practice in Offshore Engineering*. ASCE, Reston, VA. pp. 25–64.

Ohsaki, Y. 1982. Corrosion of steel piles driven in soil deposits. *Soils Found.* 22, 3, 57–76. https://doi.org/10.3208/sandf1972.22.3_57.

Park, J., and P. Towashiraporn. 2014. Rapid seismic damage assessment of railway bridges using the response-surface statistical model. *Struct. Saf.* 47, 1–12. https://doi.org/10.1016/j.strusafe.2013.10.001.

Prasad, G. and S. Banerjee, The effect of flood-induced scour on the seismic vulnerability of bridges. *J. Earthquake Eng.* 17, 6, 803–828. https://doi.org/10.1080/13632469.2013.771593.

Rampasek, L., and A. Goldenberg. 2016. Tensorflow: Biology's gateway to deep learning? *Cell Syst.* 2, 1, 12–14. https://doi.org/10.1016/j.cels.2016.01.009.

Reese, C. L., T. S. Wang, A. J. Arrellaga, J. Hendrix, and L. Vasquez. 2016. *A Group for the Analysis of a Group of Piles Subjected to Vertical and Lateral Loading (User's Manual)*. Ensoft, Austin, TX.

Samadi, S. Z., G. Sagareswar, and M. Tajiki, 2010. Comparison of general circulation models: Methodology for selecting the best GCM in Kermanshah Synoptic Station, Iran. *Int. J. Global Warm.* 2, 4, 347–365. https://doi.org/10.1504/IJGW.2010.037590.

Schlosser, F., and M. Bastick, 1991. Reinforced earth. In *Foundation Engineering Handbook*, H.-Y. Fang (ed.), Van Nostrand Reinhold, New York, pp. 778–786.

Sheffield, J., A. P. Barrett, B. Colle, D. Nelun Fernando, R. Fu, K. L. Geil, Q. Hu, J. Kinter, S. Kumar, and B. Langenbrunner, 2013. North American climate in CMIP5 experiments. Part I: Evaluation of historical simulations of continental and regional climatology. *J. Clim.* 26, 23, 9209–9245. https://doi.org/10.1175/JCLI-D-12-00592.1.

Shelhamer, E., J. Donahue, S. Karayev, J. Long, R. Girshick, S. Guadarrama, and T. Darrell. Jia, Y., 2014. Caffe: Fast feature embedding with convolutional architecture.

Proceedings of the 22nd ACM International Conference on Multimedia. Association for Computing Machinery, New York, pp. 675–678.

Shen, C. 2018. A transdisciplinary review of deep learning research and its relevance for water resources scientists. *Water Resour. Res.* 54, 11, 8558–8593. https://doi.org/10.1029/2018WR022643.

Shooman, M. L. 1968. *Probabilistic Reliability: An Engineering Approach.* McGraw-Hill, New York.

Shrestha, S., T. V. Bach, and V. P. Pandey, 2016. Climate change impacts on groundwater resources in Mekong Delta under representative concentration pathways (RCPs) scenarios. *Environ. Sci. Pol.* 61, 1–13. https://doi.org/10.1016/j.envsci.2016.03.010.

Singh, T., M. Pal, and V. K. Arora, 2019. Modeling oblique load carrying capacity of batter pile groups using neural network, random forest regression and M5 model tree. *Front. Struct. Civ. Eng.* 13, 3, 674–685. https://doi.org/10.1007/s11709-018-0505-3.

Song, L. K., G. C. Bai, C. W. Fei, and J. Wen, 2018. Reliability-based fatigue life prediction for complex structure with time-varying surrogate modeling. *Adv. Mater. Sci. Eng.* 2018, 3469465. https://doi.org/10.1155/2018/3469465.

Sun, W., B. Zheng, and W. Qian, 2016. Computer aided lung cancer diagnosis with deep learning algorithms. In *Vol. 9785 of Medical Imaging 2016: Computer-Aided Diagnosis, 97850Z.* International Society for Optics and Photonics, Bellingham, WA.

Taylor, K. E., R. J. Stouffer, and G. A. Meehl, 2012. An overview of CMIP5 and the experiment design. *Bull. Am. Meteorol. Soc.* 93, 4, 485–498. https://doi.org/10.1175/BAMS-D-11-00094.1.

The Joint Committee on Structural Safety (JCSS), 2001. *Model Code for Probabilistic.* On the 10th of November, 2019, I checked my email. https://www.jcss-lc.org/jcss-probabilistic-model-code/.

Tian, Y., K. Pei, S. Jana, and B. Ray, 2018. Deeptest: Automated testing of deep-neural-network-driven autonomous cars. In *Proceeding of 40th International Conference on Software Engineering.* Association for Computing Machinery, New York, pp. 303–314.

Turner, D. 2016. *Fragility Assessment of Bridge Superstructures under Hydrodynamic Forces.* Master's thesis, Dept. of Civil and Environmental Engineering, Colorado State Univ.

USGS. 2018. *National Water Information Service.* Accessed October 1, 2018. https://water-data.usgs.gov/nwis.

Vijayvergiya, V. N. 1977. Load-movement characteristics of piles. In *Proceedings of Ports '77 Conference.* ASCE, New York.

Wang, N. 2010. *Reliability-Based Condition Assessment of Existing Highway Bridges.* Ph.D. dissertation, School of Civil and Environmental Engineering, Georgia Institute of Technology.

Wang, S. S., L. Zhao, J. H. Yoon, P. Klotzbach, and R. R. Gillies. 2018a. Quantitative attribution of climate effects on Hurricane Harvey's extreme rainfall in Texas. *Environ. Res. Lett.* 13, 5, 054014. https://doi.org/10.1088/1748-9326/aabb85.

Wang, Z., J. E. Padgett, and L. Duenas-Osorio. 2014b. Risk-consistent calibration of load factors for the design of reinforced concrete bridges under the combined effects of earthquake and scour hazards. *Eng. Struct.* 79, 86–95. https://doi.org/10.1016/j.engstruct.2014.07.005.

Wang, Z., L. Duenas-Osorio, and J. E. Padgett. 2014a. Influence of scour effects on the seismic response of reinforced concrete bridges. Eng. Struct. 76, 202–214. https://doi.org/10.1016/j.engstruct.2014.06.026.

Wang, Z., N. Pedroni, I. Zentner, and E. Zio, 2018b. Seismic fragility analysis with artificial neural networks: Application to nuclear power plant equipment. *Eng. Struct.* 162, May, 213–225. https://doi.org/10.1016/j.engstruct.2018.02.024.

Weigel, A. P., R. Knutti, M. A. Liniger, and C. Appenzeller, 2010. Risks of model weighting in multimodel climate projections. *J. Clim.* 23, 15, 4175–4191. https://doi.org/10.1175/2010JCLI3594.1.

Westra, S., M. Thyer, M. Leonard, D. Kavetski, and M. Lambert, 2014. A strategy for diagnosing and interpreting hydrological model nonstationarity. *Water Resour. Res.* 50, 6, 5090–5113. https://doi.org/10.1002 /2013WR014719.

Winsemius, H. C., et al. 2016. Global drivers of future river flood risk. *Nat. Clim. Change* 6, 4, 381. https://doi.org/10.1038/nclimate2893.

Yip, S., C. A. Ferro, D. B. Stephenson, and E. Hawkins. 2011. A simple, coherent framework for partitioning uncertainty in climate predictions. *J. Clim.* 24, 17, 4634–4643. https://doi.org/10.1175/2011JCLI4085.1.

Zhang, Z., P. Klein, and K. Friedrich, 2002. Dynamic mechanical properties of PTFE based short carbon fibre reinforced composites: Experiment and artificial neural network prediction. *Compos. Sci. Technol.* 62, 7–8, 1001–1009. https://doi.org/10.1016/S0266-3538(02)00036-2.

15 Software Vulnerability Analysis

Rachana Kamble
Technocrats Institute of Technology

Jyoti Mishra
Gyan Ganga Institute of Technology and Sciences

Aditi Sharma
Parul University

CONTENTS

DOI: 10.1201/9780367816414-15

15.1   INTRODUCTION

Software security system is a main concern in every government and non-government organisation. Nowadays, the issue is enhancing in every fields, so are the responsibilities of the of the development organisations to improve software programs that are going to detect software-related vulnerabilities very efficiently [1]. According to the vulnerabilities are increments same as the different aspects of software program improvement security ideas should also be likewise be examination and it should be demonstrated and best practices to improve the assurance of software program structures. However, the software vulnerabilities increase day by day; according to this, our main concern is to enhance information security system. Software security devices include a massive range of complex issues that are going to be a trouble in the security system [2]. Many software program improvement corporations are working ahead and collect various software programs also developing and improving protection ideas and high-quality practices on the application program that enhance the protection of software program systems. However, the increasing software program system vulnerabilities has turned out to be one of the essential threats to the protection of data systems [3].

Machine learning is the latest technology that is used to limit various security faults in software application system; software program vulnerability evaluation is turning into the focus of records system protection technological know-how research. Machine learning concepts automatically generate expertise via massive quantities of data and by the use of the expertise for calculation. It is utilised in the discipline of textual content classification and various malicious code recognition programs. With the growing records of software program vulnerability, it has become essential to use computing devices gaining knowledge of software program vulnerability analysis or detection techniques.

15.1.1   THE FRAME OF SOFTWARE VULNERABILITY EVALUATION BASED ON MACHINE LEARNING TECHNIQUE

Machine learning strategies can substantially enhance the detection accuracy. Machine learning techniques are the way to applied on various text contents according to the categories also applied to detect various malicious code that resides on software contents. In the manner of textual content classification, it used Salton vector area model that effectively specific files with the series of words, and after then embed them into vector space, the machine learning techniques are very used that extract points that is specifically generate classification model. In the detection of malicious code, unique detection signatures need to be written, which can healthily detect infected vulnerabilities on host's network.

The applications containing vulnerabilities consist of a big vocabulary block of code, which has a complicated relationship. Therefore, software evaluation and function extraction are used first and then machine learning strategies are used to acquire vulnerability evaluation and localisation. According this process, software program vulnerability evaluation framework-based totally on machine learning can be divided into four steps as follows:

 i. Application analysis
 ii. Function extraction
 iii. Computer learning
 iv. Vulnerability location.

as nicely as training and evaluation stages, as proven below.

At the duration of training stage, it focused on program analysis, feature extraction it processes the greater security application code and the vulnerable software code are in the training set, after the application evaluation and function extraction, the end result of these two steps have been enter into computing device learning algorithms and acquired the classifier of software program vulnerability analysis, shown in Figure 15.1a. In the evaluation stage, the software code was once analysed and characteristic extraction first, thru the classifier to determine if it consists of the vulnerability, and come across the vulnerability role in accordance with the features of vulnerability that has been proven in Figure 15.1b.

This framework is based on the current vulnerability evaluation techniques that are used with the machine learning techniques that can be categorised by using three approaches as shown in the diagram below.

Software vulnerabilities can be explained as weaknesses or faults present in any kind of software or application. Inappropriate testing and manual code reviews are not at all a good option, and they cannot always find each and every vulnerability. Basically, vulnerabilities can decrease the performance and security of the application software. They will also allow unauthenticated attackers or unauthorised users to exploit or gain access to particular products and data. So it is mandatory to be aware all the top 10 most common vulnerabilities for detecting software vulnerabilities. There are various techniques to detect software vulnerabilities, and by using them, in-built software vulnerabilities can be easily identified and prevented. A very important thing is that we should be aware of or have knowledge of several vulnerabilities. Here, we go through definitions of vulnerabilities and deliver a list of the top 10 software vulnerabilities and guidelines on how to prevent software vulnerabilities.

15.2   TOP 10 MOST COMMON SOFTWARE VULNERABILITIES

According to the OWASP, there are ten most important vulnerabilities (Figure 15.2).

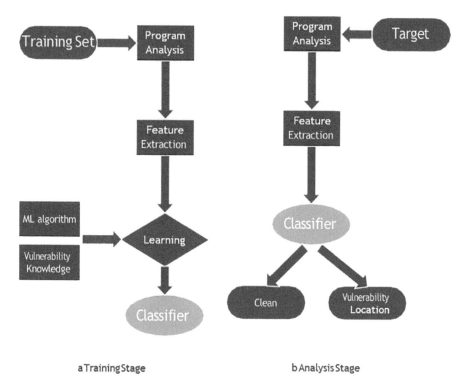

FIGURE 15.1 A framework of software vulnerability analysis based on machine learning technique. (a) Training Stage (b) Analysis Stage.

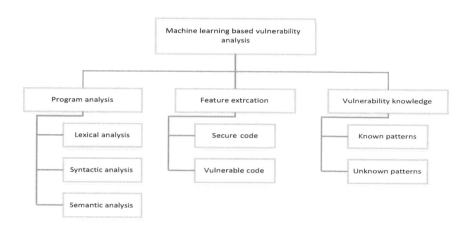

FIGURE 15.2 Methods classification (machine learning-based vulnerability analysis).

15.2.1 BROKEN ACCESS CONTROL

User limitations must be suitably enforced. At some situations, if it is broken, in that particular case, it can generate a software vulnerability. Unauthorised attackers can exploit or gain that vulnerability.

15.2.2 CRYPTOGRAPHIC FAILURES

Users are having lots of sensitive data or information such as addresses, passwords and bank account details. All these things should be properly protected. If it is not, in that case unauthorised attackers can take benefit of the vulnerabilities to gain access to the personal credentials of the users.

15.2.3 INJECTION

Injection attacks occur when untrusted data or some kind of Trojan program is sent as portion of a command or sub-query language. Then this attack is executed into the targeted system, and abnormal activities are activated. An attack can also provide unauthorised attackers admittance to protected data.

15.2.4 INSECURE DESIGN

Insecure design refers to risks related to design flaws, which often include the lack of at least one of the following:

- Threat modelling
- Secure design patterns
- Secure design principles
- Reference architecture.

15.2.5 SECURITY MISCONFIGURATION

Security misconfigurations commonly give the following results:

- Insecure or unconfident default configurations
- Imperfect or impromptu configurations
- Access to open cloud storage
- Misconfigured HTTP headers
- Wordy error messages that comprise sensitive information.

15.2.6 VULNERABLE AND OUTDATED COMPONENTS

Components are the groupings of libraries, outlines and other software modules. Regularly, the components run on the same rights as per the application. In case if a component originates as a vulnerability, it can be exploited by an unauthorised attacker. This leads to serious data loss or will result in the server being hacked.

15.2.7 IDENTIFICATION AND AUTHENTICATION FAILURES

In software security authentication and session management programs, applications and their credentials have to be applied correctly. If there is any lack or mistakes in the functionalities, it generates a software vulnerability that can be oppressed by unauthorised attackers to gain admittance to individuals' data.

15.2.8 SOFTWARE AND DATA INTEGRITY FAILURES

Software and records integrity screw-ups refer to assumptions made about software program updates, imperative data and CI/CD pipelines barring verifying integrity. In addition, deserialisation flaws regularly result in remote code execution (RCE is a type of software vulnerability). This allows untrustworthy sellers to execute restore, injection and privilege growth attacks.

15.2.9 SECURITY LOGGING AND MONITORING FAILURES

Sometimes, the lack of logging and monitoring processes are very unsafe, which leaves users' data vulnerable to interference, removal or even demolition.

15.2.10 ON THE SERVER-SIDE REQUEST/RESPONSE FORGERY ATTACK

Server-side request/response forgery attacks are a most common hacking strategy that provides access to the server and wanted to do some unauthorised work on that it also refers to information or any useful data that recommends a noticeably very lower occurrence rate that is showing some common testing coverage, and several ranking for elaborate various kinds of impact potential.

15.3 STEPS TO PREVENT SOFTWARE VULNERABILITIES

There are basically three most advantageous software programs to prevent software program weaknesses.

15.3.1 CREATE SOFTWARE DESIGN REQUIREMENTS

Software program design requirements that define and implement invulnerable coding principles have been established. This has to comprise the use of a secure coding standard.

15.3.2 USE A CODING STANDARD

Coding necessities are, for example, OWASP, CWE and CERT, which are going to prevent, detect and remove vulnerabilities. That is going to enforcing a coding general is effortless when user or candidate use a SAST device such as Klocwork. Basically, Klocwork classifies safety defects and vulnerabilities although the code is going to be written.

15.3.3 Test Your Software

It is important that that the user check the software program in a very early and regular basis. These assistances make sure that vulnerabilities are experiential and removed as quickly as possible depending on the situation. A most common approach used by the programmer is called static code analyser, such as Clockwork. The clockwork is a phase of the software program that checks out process execution.

15.3.4 Vulnerability Assessment Tools

Vulnerability assessments are frequently carried out to assist guarantee groups are included from normal vulnerabilities (low placing fruit). Vulnerability scanning, evaluation and administration all share an integral cybersecurity principle: The horrific guys can't get in if they don't have a way. To that end, an imperative IT safety exercise is to scan for vulnerabilities and then patch them, usually by means of a patch administration system.

Vulnerability scanning equipment can make that technique less complicated with the aid of discovering and even patching vulnerabilities for you, decreasing the burden on protection group of workers and operations centres. Vulnerability scanners notice and classify device weaknesses to prioritise fixes and so often predict the effectiveness of countermeasures. Scans can be carried out by way of the IT branch or by a provider. Typically, the scan compares the important points of the target attack surface to a database of data about recognised safety holes in offerings and ports, as properly as anomalies in packet construction, and paths that may additionally exist to exploitable applications or scripts.

Some scans are carried out by means of logging in as an approved person, while others are carried out externally and strive to find holes that can also be exploitable via these running outside the network. Vulnerability scanning must no longer be burdened with penetration testing, which is about exploiting vulnerabilities alternatively than indicating the place conceivable vulnerabilities may lie. Vulnerability administration is a broader product that contains vulnerability scanning capabilities, and a complementary technological know-how is breach and attack simulation, which permits for non-stop automatic vulnerability assessment.

Depending on the areas of the infrastructure under assessment, a vulnerability evaluation can be categorised into three wide types.

External Scans: Scanning these factors of the IT ecosystem that without delay face the web and are reachable to exterior users, for instance, ports, networks, websites, apps and different structures used through exterior customers or customers.
Internal Scans: Finding loopholes in the interior community of a business enterprise (do not cover exterior scans) that might also harm the enterprise network.
Environmental Scans: Environmental vulnerability scans focus on precise operational science of an organisation, such as cloud services, IoT and cell devices.

15.4   TOP 10 VULNERABILITY EVALUATION TOOLS

Nikto2: It is an open-source vulnerability scanning evaluation software program pivoting on net utility security. Nikto2 can notice around 6700 malicious archives inflicting a risk to net servers disclosing out-of-date servers [4-5]. Nikto2 watches on server configuration problems by means of performing net server scans within a quick time. Nikto2 does now not have any expedients to vulnerabilities detected and additionally does now not supply chance evaluation features. Nikto2 is up to date now and then for overlaying broader vulnerabilities [6-7].

Netsparker: A device with Internet software vulnerability embedded with an automatic characteristic for detecting vulnerabilities. This device is expert in assessing vulnerabilities in a number of net purposes within a particular time [8-9].

OpenVAS: A sturdy vulnerability scanning device aiding large-scale scans for organisations. This device is really useful in detecting vulnerabilities in the net utility or net servers and databases, running systems, networks and digital machines [10-11]. OpenVAS gets day-by-day updates widening the vulnerability detection coverage. It is beneficial in chance evaluation recommending expedients for detecting vulnerabilities [12-13].

W3AF: This is an untethered and open-source device additionally acknowledged as web-application-attack and framework. It is an open-source evaluation device for Internet applications. It types a framework for securing Internet purposes by using detecting and making use of the vulnerabilities. An undemanding device with points of vulnerability scanning, W3AF has extra amenities for penetration checking. Furthermore, W3AF has a different collection of vulnerabilities. This device is especially really helpful for domains that are at stake often with vulnerabilities that are currently identified.

Arachni: This is an unwavering vulnerability device for Internet purposes and is many times updated. This has a broader insurance of vulnerabilities and has selections for threat evaluation recommending hints and counter element for the vulnerabilities detected.

Acunetix: This is a paid Internet evaluation software safety device that is open source with many purposes. This device has a broader vulnerability scanning range, with over 6500 vulnerabilities. It can notice community vulnerabilities along net applications. It is a device that permits automating our assessment. This is excellent for large-scale companies as it can manoeuvre various devices.

Nmap: It is a famous and free open-source community evaluation device among many protection professionals. Nmap maps with the aid of inspecting hosts in the community for figuring out the working systems. This characteristic is beneficial in discovering vulnerabilities in single or more than one network.

OpenSCAP: It is a structured equipment that helps in vulnerability scanning, assessment and measurement, forming a safety measure. It is a neighbourhood-developed device assisting Linux platforms. OpenSCAP framework presents power to the vulnerability evaluation on net applications, servers, databases, working systems, networks and digital machines. They additionally investigate danger and counteract threats.

GoLismero: It is an unpaid open-source device for assessing vulnerability. It is a device specialised in detecting vulnerabilities on net functions and networks [14-15]. It is a device of comfort performing with the output furnished through different vulnerability equipment such as OpenVAS that combines output with the feedback. It additionally covers database and community vulnerabilities [16-17].

Intruder: It is a paid device for vulnerability evaluation designed to determine cloud-based storage. Intruder software program assesses the vulnerability immediately after its release [18-19]. It has computerised scanning points that consistently video display units for vulnerability, with the aid of presenting high-quality reports [20-21].

15.5 VULNERABILITY ASSESSMENT AND PENETRATION TESTING

Vulnerability evaluation is a technique in which the IT structures such as computer systems and networks and software programs such as operating systems and utility software program are scanned in order to discover the presence of regarded and unknown vulnerabilities.

As many as 80% of Internet websites have vulnerabilities that may lead to the theft of sensitive company records such as savings card data and purchaser lists.

Hackers are concentrating their efforts on web-based purposes – buying carts, forms, login pages, dynamic content, etc. Accessible throughout the world, insecure Internet functions grant easy get right of entry to backend company databases.

VAPT can be carried out in the following nine-step process.

15.5.1 SCOPE

While performing assessments and tests, the scope of the task desires to be absolutely defined. The scope is based totally on the belongings to be tested. The following are the three viable scopes that exist.

15.5.2 BLACK BOX TESTING

Testing from an exterior community with no prior information of the inner networks and systems.

15.5.3 GREY BOX TESTING

Testing from an exterior or interior network, with the know-how of the interior networks and systems. This is commonly a mixture of black container checking out and white container testing.

15.5.4   WHITE BOX TESTING

Performing the check from within the community with the understanding of the community structure and the systems. This is additionally referred to as inside testing.

15.5.5   INFORMATION GATHERING

The procedure of records gathering is to attain as plenty records as viable about the IT surroundings such as networks, IP addresses, and running gadget version. This is relevant to all the three sorts of scope as mentioned earlier.

15.5.6   VULNERABILITY DETECTION

In this process, equipment such as vulnerability scanners is used, and vulnerabilities are recognised in the IT surroundings through way of scanning.

15.5.7   INFORMATION ANALYSIS AND PLANNING

This procedure is used to analyse the recognised vulnerabilities, mixed with the facts gathered about the IT environment, to devise a diagram for penetrating into the community and system.

15.5.8   PENETRATION TESTING

In this process, the goal structures are attacked and penetrated through the usage of the diagram devised in the process before.

15.5.9   PRIVILEGE ESCALATION

After profitable penetration into the system, this procedure is used to perceive and improve getting admission to attain greater privileges, such as root get entry or administrative get entry to the system.

15.6   RESULTS ANALYSIS

This procedure is beneficial for performing a root reason evaluation as an end result of a profitable compromise to the gadget main to penetration and devise appropriate hints in order to make the machine invulnerable via plugging the holes in the system.

15.7   REPORTING

Every one of the discoveries not entirely set in stone in that frame of mind of the weakness assessment and infiltration evaluating technique need to be reported, close by with the proposals, to create the evaluating report to the organization for proper activities.

15.7.1 Clean-Up Activity of Vulnerability

Vulnerability evaluation and penetration testing includes compromising the system, and at some point, of the process, some of the documents may additionally be altered. This method ensures that the machine is delivered returned to the unique state, before the testing, with the aid of cleansing up (restoring) the facts and documents used in the goal machines.

15.8 CONCLUSIONS

Nowadays, software vulnerabilities strategies are improving day by day. Whenever it founds new vulnerabilities according to that the research work is going on to analysis and detect it. If we talk about traditional strategies that are truly based on observed vulnerabilities or the history of vulnerabilities and that are going to confined various guidelines of previous vulnerability results or detections scenarios. When the new software program structures occur again and again, the current strategies conduct the report, such as false positives and false negatives accordingly. Recently, the machine learning technologies have been utilised to analyse software program vulnerability. They are a very popular evaluation approach and have lots of dynamic fundamentals and mechanisms that effectively finds out new and upcoming software vulnerabilities. Also, they are able to enhance the effectivity of software program vulnerability evaluation significantly.

In this book chapter, an overview of familiar works that use computing devices to analyse the software program vulnerabilities has been provided. It proposed a software program vulnerability evaluation framework that is totally based on machine learning techniques and vulnerabilities analysis tools. These tools are categorised and are very helpful for detecting numerous vulnerabilities that tools are applied with machine learning technologies.

REFERENCES

[1] Wu, S., et al. Software vulnerability analysis technology progress. *Journal of Tsinghua University (Natural Science)* 10 (2012): 1309–1319.

[2] Balzarotti, D., et al. Saner: Composing static and dynamic analysis to validate sanitization in web applications. *Security and Privacy, 2008. SP 2008. IEEE Symposium on IEEE*, 2008.

[3] Sebastiani, F. Machine learning in automated text categorization. *ACM Computing Surveys (CSUR)* 34.1 (2002): 1–47.

[4] Shabtai, A., et al. Detection of malicious code by applying machine learning classifiers on static features: A state-of-the-art survey. *Information Security Technical Report* 14.1 (2009): 16–29.

[5] Yamaguchi, F., L. Felix, and K. Rieck. Vulnerability extrapolation: Assisted discovery of vulnerabilities using machine learning. *Proceedings of the 5th USENIX Conference on Offensive Technologies. USENIX Association*, 2011.

[6] Yamaguchi, F., M. Lottmann, and K. Rieck. Generalized vulnerability extrapolation using abstract syntax trees. *Proceedings of the 28th Annual Computer Security Applications Conference. ACM*, 2012.

[7] Yamaguchi, F., et al. Chucky: Exposing missing checks in source code for vulnerability discovery. *Proceedings of the 2013 ACM SIGSAC conference on Computer & Communications Security. ACM*, 2013.

[8] Yamaguchi, F., et al. Modeling and discovering vulnerabilities with code property graphs. *Security and Privacy (SP), 2014 IEEE Symposium on. IEEE*, 2014.

[9] Yamaguchi, F., et al. Automatic Inference of Search Patterns for Taint-Style Vulnerabilities, *IEEE Symposium on Security and Privacy*, 797–812. IEEE, 2015.

[10] Zhang, S., D. Caragea, and X. Ou. An empirical study on using the national vulnerability database to predict software vulnerabilities. *Database and Expert Systems Applications*. Springer, Berlin Heidelberg, 2011.

[11] Grieco, G., et al. Toward Large-Scale Vulnerability Discovery Using Machine Learning. *Proceedings of the Sixth ACM on Conference on Data and Application Security and Privacy, CODASPY 2016*. ACM, New Orleans, LA, USA, March 9-11, 2016 : 85–96.

[12] Wang, Y., Y. Wang, and J. Ren. Software vulnerabilities detection using rapid density-based clustering. *Journal of Computational Information Systems* 8.14(2011): 3295–3302.

[13] Cheng, H., et al. Identifying bug signatures using discriminative graph mining. *Proceedings of the Eighteenth International Symposium on Software Testing and Analysis. ACM*, 2009.

[14] Wijayasekara, D., et al. Mining bug databases for unidentified software vulnerabilities. *Human System Interactions (HSI), 2012 5th International Conference on. IEEE*, 2012.

[15] Neuhaus, S., et al. Predicting vulnerable software components. *Proceedings of the 14th ACM Conference on Computer and Communications Security. ACM*, 2007.

[16] Mokhov, S., J. Paquet, and M. Debbabi. MARFCAT: Fast code analysis for defects and vulnerabilities. *Software Analytics (SWAN), 2015 IEEE 1st International Workshop on. IEEE*, 2015.

[17] Almorsy, M., J. Grundy, and A. S. Ibrahim. Supporting automated vulnerability analysis using formalized vulnerability signatures. *Proceedings of the 27th IEEE/ACM International Conference on Automated Software Engineering. ACM*, 2012.

[18] Medeiros, I., N. F. Neves, and M. Correia. Automatic detection and correction of web application vulnerabilities using data mining to predict false positives. *Proceedings of the 23rd International Conference on World Wide Web. ACM*, 2014.

[19] Shar, L. K., and H. B. Kuan Tan. Predicting SQL injection and cross site scripting vulnerabilities through mining input sanitization patterns. *Information and Software Technology* 55.10(2013): 1767–1780.

[20] Liu, H., et al. PF-miner: A new paired functions mining method for Android kernel in error paths. *Computer Software and Applications Conference (COMPSAC), 2014 IEEE 38th Annual. IEEE*, 2014.

[21] Liu, H.-Q., et al. BP-Miner: mining paired functions from the binary code of drivers for error handling. *2014 21st Asia-Pacific Software Engineering Conference* 58(2014): 5A4–5A9.

Index